Business Guides on the Go

"Business Guides on the Go" presents cutting-edge insights from practice on particular topics within the fields of business, management, and finance. Written by practitioners and experts in a concise and accessible form the series provides professionals with a general understanding and a first practical approach to latest developments in business strategy, leadership, operations, HR management, innovation and technology management, marketing or digitalization. Students of business administration or management will also benefit from these practical guides for their future occupation/careers.

These Guides suit the needs of today's fast reader.

More information about this series at
https://link.springer.com/bookseries/16836

Minh Trang Rausch-Phan
Patrick Siegfried

# Sustainable Supply Chain Management

## Learning from the German Automotive Industry

Minh Trang Rausch-Phan
ISM International School of
Management GmbH
Frankfurt am Main, Germany

Patrick Siegfried
ISM International School of
Management GmbH
Frankfurt am Main, Germany

ISSN 2731-4758 ISSN 2731-4766 (electronic)
Business Guides on the Go
ISBN 978-3-030-92155-2     ISBN 978-3-030-92156-9  (eBook)
https://doi.org/10.1007/978-3-030-92156-9

© The Editor(s) (if applicable) and The Author(s), under exclusive licence to Springer Nature Switzerland AG 2022

This work is subject to copyright. All rights are solely and exclusively licensed by the Publisher, whether the whole or part of the material is concerned, specifically the rights of translation, reprinting, reuse of illustrations, recitation, broadcasting, reproduction on microfilms or in any other physical way, and transmission or information storage and retrieval, electronic adaptation, computer software, or by similar or dissimilar methodology now known or hereafter developed.

The use of general descriptive names, registered names, trademarks, service marks, etc. in this publication does not imply, even in the absence of a specific statement, that such names are exempt from the relevant protective laws and regulations and therefore free for general use.

The publisher, the authors and the editors are safe to assume that the advice and information in this book are believed to be true and accurate at the date of publication. Neither the publisher nor the authors or the editors give a warranty, expressed or implied, with respect to the material contained herein or for any errors or omissions that may have been made. The publisher remains neutral with regard to jurisdictional claims in published maps and institutional affiliations.

This Springer imprint is published by the registered company Springer Nature Switzerland AG.
The registered company address is: Gewerbestrasse 11, 6330 Cham, Switzerland

# Preface

This book researches current causes and effects of implementing sustainable supply chain management (SSCM) as well as green supply chain management (GSCM) strategies in the automotive industry. It includes a literature review on SSCM and GSCM and comprises the advantages of sustainable development concepts as well as factors causing the implementation of SSCM like buyers' behaviour, governmental regulations and competitiveness. Sustainability consists of economic, environmental and social aspects. SSCM is the management of the flow of materials and goods with the purpose of a minimisation of harmful impacts on the environment while still creating economic advantages and contributing to social responsibilities. This research focuses on discussing the green supply chain which includes economic and environmental performances but leaves out the social aspect.

Consequently, the current situation of SSCM development is analysed, particularly in the automotive industry. Challenges, barriers, successes and benefits that automotive companies obtain from implementing GSCM are shown. Through case studies on the leading German car manufacturers VW, BMW and Daimler, the necessary activities of these companies to implement green development in the entire supply chain including green supplier selection, green materials, green transportation and reverse logistics are defined. Moreover, a benchmark with companies from Asian markets like Toyota from Japan and Geely from China is performed.

Additionally the customers' interest and expectations on the implementation of SSCM in the automotive industry are highlighted by showing some of the results of a questionnaire that was given to possible vehicle users. The findings indicate that the interest of customers in SSCM increased within the last years combined with growing intentions of purchasing green vehicles. For the majority of the customers, it is important that the entire lifecycle of their vehicle is sustainable. The implementation of SSCM enhances the trust of customers in companies. They are willing to pay higher prices for vehicles which are assembled through a sustainable supply chain. Therefore, a higher economic value is created for companies with a SSCM implementation.

Based on the theoretical findings from the literature review and empirical findings from the customer survey, this research suggests that automotive companies should invest and pay more attention on practising SSCM effectively in order to achieve greater improvements in their environmental performances and reach higher economic benefits in the future. It is necessary for automotive companies to establish overall directions and useful guidelines for all stakeholders and actors along the supply chain. Training courses and capacity building programmes help employees and managers to recognise the high benefits of approaching sustainability concepts. In the future development of SSCM, customers still play important roles for companies' decision that directly affect the business' revenues and benefits. Therefore, automotive manufacturers need to base their sustainable strategies depending on customers' expectations and requirements. To gain more customer attention, it is recommended for automotive companies to develop a sustainable supply chain label and place it on their new vehicles. This label could give more details about the sustainable activities along the entire value chain of the vehicle to possible buyers. Higher customer loyalty and a better environmentally friendly brand image can be established which could result in long-term economic benefits for automotive manufacturers through higher sales and revenues.

Frankfurt am Main, Germany
Minh Trang Rausch-Phan
Patrick Siegfried

# Contents

1 **Introduction of Sustainable Supply Chain Management: Learning from the German Automotive Industry**    1
   1.1  Introduction of Sustainable Supply Chain Management Implementation in the German Automotive Industry    1
   References    5

2 **Traditional Supply Chain Management**    7
   2.1  Definition of Traditional Supply Chain Management    7
   2.2  Traditional Supply Chain Management in the Automotive Industry    10
   References    13

3 **Sustainable Supply Chain Management**    17
   3.1  Sustainable Development    17
      3.1.1  Definition and Advantages of Sustainability    17
      3.1.2  Sustainability Framework: Triple Bottom Line (TBL)    19
      3.1.3  Regulations for Sustainable Development    20
   3.2  Definition of Sustainable Supply Chain Management    21
   3.3  Driving Force Factors    24
      3.3.1  Customers' Behaviour    25
      3.3.2  Governmental Regulations    26

|  |  | 3.3.3 | Competitors | 27 |
|---|---|---|---|---|
|  |  | 3.3.4 | Innovative Technology Development in Sustainable Supply Chain | 28 |
|  | 3.4 | Performance Measures for Sustainable Supply Chain Management | | 29 |
|  |  | 3.4.1 | Economic Performance | 30 |
|  |  | 3.4.2 | Environmental Performance | 31 |
|  | 3.5 | Causes for Implementing Sustainable Supply Chain Management in the Automotive Industry | | 32 |
|  |  | 3.5.1 | Growing Competitive Markets | 33 |
|  |  | 3.5.2 | Green Consumers | 35 |
|  |  | 3.5.3 | Environmental Global Laws | 38 |
|  | References | | | 39 |
| **4** | **Green Supply Chain Management** | | | **47** |
|  | 4.1 | Introduction to Green Supply Chain Management | | 47 |
|  | 4.2 | Principles of Implementing Green Supply Chain Management | | 49 |
|  |  | 4.2.1 | Green Suppliers Selection | 50 |
|  |  | 4.2.2 | Green Product Design | 51 |
|  |  | 4.2.3 | Green Material Purchasing | 51 |
|  |  | 4.2.4 | Green Manufacturing | 52 |
|  |  | 4.2.5 | Green Distribution | 53 |
|  |  | 4.2.6 | Reverse Logistics | 54 |
|  | 4.3 | Benefits of Applying Green Supply Chain Management in the Automotive Industry | | 54 |
|  |  | 4.3.1 | Benefits from Environmental Performance | 55 |
|  |  | 4.3.2 | Benefits from the Economic Performance | 61 |
|  | 4.4 | Barriers and Challenges of Green Supply Chain Management in the Automotive Industry | | 62 |
|  | 4.5 | Case Studies: German Automotive Companies Using Green Supply Chain Management | | 65 |
|  |  | 4.5.1 | German Automotive Industry | 65 |
|  |  | 4.5.2 | Volkswagen | 67 |
|  |  | 4.5.3 | BMW | 71 |

|  |  | | |
|---|---|---|---|
| | 4.5.4 | Daimler | 74 |
| | 4.5.5 | Benchmark Against Automotive Manufacturers in Asian Countries | 77 |
| References | | | 82 |

## 5 Scenarios and Concepts for the Future Development — 91
5.1 Partnerships Between the Manufacturers — 91
5.2 Extending Environmental Initiatives Throughout Supply Chain Actors — 92
5.3 Customer Orientation as a High Influence Factor — 93
5.4 Providing Information to Customers Through SSC Labelling — 95
References — 97

## 6 Conclusion of Sustainable Supply Chain Management: Learning from the German Automotive Industry — 99
References — 102

## Appendix — 103

# Abbreviations

| | |
|---|---|
| 3PL | Third-Party Logistics |
| ACC | American Chemistry Council |
| ACEA | European Automobile Manufacturers' Association |
| APICS | American Production and Inventory Control Society |
| BEV | Battery Electric Vehicle |
| BRIC | Brazil, Russia, India, China |
| CAGR | Compound Annual Growth Rate |
| CDP | Carbon Disclosure Project |
| CFRP | Carbon Fibre Reinforced Plastic |
| Eco-VAS | Eco-Vehicle Assessment System |
| ELV | End-of-Life Vehicles |
| EMAS | Eco-Management and Audit Scheme |
| EMS | Environmental Management System |
| EU | European Union |
| EU ETS | EU Emission Trading System |
| EuP | Energy-Using Products |
| EV | Electric Vehicle |
| FCV | Fuel Cell Vehicles |
| GDP | Gross Domestic Product |
| GHG | Greenhouse Gas |
| GSCM | Green Supply Chain Management |
| HEV | Hybrid Electric Vehicle |
| IEA | International Energy Agency |

| | |
|---|---|
| ISO | International Organization for Standardization |
| NEV | New Energy Vehicle |
| $NO_X$ | Nitrogen Oxide |
| NPE | National Platform for Electric Mobility |
| OEM | Original Equipment Manufacturer |
| PHEV | Plug-in Hybrid Electric Vehicle |
| Pkw-EnVKV | Pkw-Energieverbrauchskennzeichnungsverordnung |
| RoHS | Restriction of Hazardous Substances |
| SCM | Supply Chain Management |
| SSC | Sustainable Supply Chain |
| SSCM | Sustainable Supply Chain Management |
| TBL | Triple Bottom Line |
| UNFCCC | United Nations Framework Convention on Climate Change |
| USA | United States of America |
| VDA | German Association of the Automotive Industry |
| VW | Volkswagen |
| WEEE | Waste Electrical and Electronic Equipment |
| WLTP | Worldwide Harmonised Light Vehicles Test Procedure |

# List of Figures

| | | |
|---|---|---|
| Fig. 2.1 | The activities of supply chain management (own illustration based on Jammernegg et al., 2009) | 9 |
| Fig. 2.2 | Levels of supply chain complexity (Díaz, 2006) | 12 |
| Fig. 3.1 | Sustainable supply chain management (Carter & Rogers, 2008) | 22 |
| Fig. 3.2 | Supply chain sustainable framework (Brandenburg & Rebs, 2015) | 24 |
| Fig. 3.3 | Market capitalisation of automotive brands (Richter, 2020) | 33 |
| Fig. 3.4 | Global electric car fleet (Virta, 2020) | 36 |
| Fig. 3.5 | Global BEV and PHEV deliveries (Virta, 2020) | 37 |
| Fig. 4.1 | Benefits of green supply chain management (Sanket Tonape, 2013) | 48 |
| Fig. 4.2 | Green supply chain management process (own illustration based on Diabat & Govindan, 2011) | 49 |
| Fig. 4.3 | The hierarchy of options for the treatment of waste during the manufacture of vehicles (own illustration based on Gaudillat et al., 2017) | 58 |
| Fig. 4.4 | Primary energy requirement of different vehicle engine types (own illustration based on Volkswagen AG, 2020b) | 68 |
| Fig. 4.5 | $CO_2$ emissions per vehicle produced (BMW Group, 2020) | 74 |
| Fig. 4.6 | Number of passenger cars sold in the Asia Pacific region 2019, by country or region (own illustration based on Moore, 2020) | 77 |
| Fig. 5.1 | Importance of buying decisions | 94 |

| | | |
|---|---|---|
| Fig. 5.2 | Finding information about sustainability | 96 |
| Fig. 5.3 | Accepting higher prices | 96 |
| Fig. A.1 | Participants' gender | 104 |
| Fig. A.2 | Participants' age | 105 |
| Fig. A.3 | Participants' highest completed degree | 105 |
| Fig. A.4 | Participants' employment status | 106 |
| Fig. A.5 | Participants' current car engine | 106 |
| Fig. A.6 | Results of question six | 107 |
| Fig. A.7 | Results of question six stacked by results of question seven | 108 |
| Fig. A.8 | Results of question nine | 108 |
| Fig. A.9 | Results of question 10 | 109 |
| Fig. A.10 | Results of question 11 | 109 |
| Fig. A.11 | Results of question 10 clustered by results of question seven | 110 |
| Fig. A.12 | Results of question 21 | 111 |
| Fig. A.13 | Results of question 17 | 111 |
| Fig. A.14 | Results of question 17 clustered by results of question 18 | 112 |

# 1

# Introduction of Sustainable Supply Chain Management: Learning from the German Automotive Industry

## 1.1 Introduction of Sustainable Supply Chain Management Implementation in the German Automotive Industry

One of the most discussed and most concerning topics in the recent years for human society is global warming. Scientists warn that human activities like burning fossil fuels and deforestation can cause rising global surface temperatures. Throughout history and parallel to the industrial revolution, the climate changed dramatically with effects like rising global sea level, melting ice sheets and a rise in natural disasters (WWF, 2019). Therefore, environmental issues are becoming more urgent and the awareness of protecting the human environment is regarded as the most important activity nowadays. Research by the American Chemistry Council (ACC) has indicated the growing number of shoppers who buy more sustainable products. That shows the shift in the consumers' buying behaviour (Accenture, 2019).

Different from the past, when pricing was the first decision criterion, consumers are actively choosing more environmental-friendly products and are willing to pay extra money to sustainable companies in the recent

© The Author(s), under exclusive license to Springer Nature Switzerland AG 2022
M. T. Rausch-Phan, P. Siegfried, *Sustainable Supply Chain Management*, Business Guides on the Go, https://doi.org/10.1007/978-3-030-92156-9_1

years (Martins, 2019; Siegfried, 2020). The survey "International Electric-Vehicle Consumer Survey" by AlixPartners has shown that consumers' interests in electric vehicles are increasing. According to AlixPartners, 50% of the surveyed consumers are interested in owning a battery electric vehicle (BEV) and 28% would purchase a BEV as their next vehicle (Bastin et al., 2019; Siegfried, 2021a). Hence, the trend of choosing green mobilities reinforces manufacturers in using sustainable practices and performances in their manufacturing and their business' targets. Sustainability development turns into a crucial strategy for companies in all industries and the automotive industry is no exception to this rule.

The automotive industry contributes high revenues to the world's economy and consists of many large car manufacturing companies and suppliers which are building and selling vehicles as well as purchasing materials and spare parts from a variety of destinations globally (Adams, 1981; Siegfried, 2021b). The growth of the automotive industry leads to an increase of $CO_2$ emissions based on road transport. The transport sector alone is responsible for 14% of the global greenhouse gas emissions (PricewaterhouseCoopers, 2007). On top, the automotive industry processes high numbers of components made from rubbers, plastics or steels which are problematic to recycle.

To reduce harmful impacts on the environment, governments have imposed stricter environmental regulations. The $CO_2$ emissions of new registered cars in the year 2030 need to be reduced by 37.5% compared to the current limits (European Council, 2019). Besides there are more waste management regulations like the "Waste Management Licensing Regulations 1994" to control the companies' material sourcing (Elghali et al., 2004).

In order to deal with governmental regulations on reducing $CO_2$ emissions and as a response to changes in consumers' behaviour on buying green vehicles, the automotive manufacturers have researched and integrated the sustainability concept into their entire supply chain management. The traditional supply chain management prioritises economic value as the main objective despite the harms on the environment caused by activities in the supply chain (Ernst & Sailer, 2015). Different from that, the sustainable supply chain management (SSCM) concerns the three aspects: economic, social and environmental which all need to be considered to satisfy the stakeholders' and customers' requirements and to grow the business (Morana, 2013). SSCM manages the flow of materials and goods with the purpose of minimisation of harmful impacts on the environment and still creates economic advantages and contributes to social responsibilities at the same time (Hu & Hsu, 2010). Many automakers see sustainability as one of the most considerable topics throughout the automotive industry because it leads to competitive advantages and improves profitability (McCrea, 2019).

Understanding the importance of sustainability, many authors have published their research in the field of SSCM since the 2010s, most of them in the general fields. Morana (2013) examined SSCM in an overall theoretical observation of the three aspects: economic, social and environmental. Rajeev et al. (2017) covered comprehensive reviews of 59 papers researching different issues of SSCM between 2000 and 2015. Recently, the sustainability in the automotive industry was researched too. For example, there is a study on designing guidelines and process-oriented views on SSCM in the automotive industry (Masoumi et al., 2019).

Resulting from the generalisation of previous theoretical frameworks and researches on SSCM, this study examines not only findings on sustainable supply chain management of automotive manufacturers but also determines driving factors forcing companies to make SSCM decisions. Possible driving factors are competitors, consumer requirements and governmental regulations. In particular, this study highlights how the consumers' behaviour impacts on the decision of manufacturers to implement SSCM. Besides, this paper focuses on two substantial aspects of SSCM which are the economic and environmental performances. Practical experiences show that many companies are still facing challenges with how to operate with environmental production processes and to enhance economic profits at the same time (Gifford, 1997; Siegfried, 2015a, 2015b).

In comparison to SSCM, Green Supply Chain Management (GSCM) just includes the two aspects economic and environmental and leaves out the social aspect. It will be highlighted in this study as an approach of automakers for reaching environmental sustainability in accordance with changes of customers' preferences on green vehicles and new environmental regulations. GSCM covers stages of upstream and downstream within the supply chain from product design, supplier selection, material purchasing, manufacturing processes, delivery of final products to end-users and disposal at the end of a product's life cycle (Emmett & Sood, 2010). Furthermore, this paper observes how German automotive producers have implemented SSCM in their system in comparison with other original equipment manufacturers (OEMs) from Asian markets like China and Japan.

The overall aim of this study is to analyse current effects and challenges of implementing SSCM as well as GSCM strategies in the automotive industry. Three big German automotive manufacturers, Volkswagen, BMW and Daimler, are selected to benchmark them compared to sustainable practices from other global rivals from the Asian markets. After all, due to the globalisation and growth of the automotive industry moving towards zero emissions by using renewable energies, this paper will investigate how customers' requirements may affect and pressure automotive OEMs for the forthcoming utilisation of sustainable supply chain management (SSCM) by using a questionnaire survey on car drivers (see Appendix). Finally, based on the findings in the literature and of customers' perceptions in the survey, recommendations for future amendments in the automotive industry will be suggested.

The objectives of this research are defined as follows:

- Review the theoretical understanding of the traditional and sustainable supply chain and determine differences between them.
- Define sustainable strategies based on the triple bottom line (TBL) framework through three dimensions: economic, social and environmental and the advantages of sustainable development.
- Determine the driving factors which influence sustainable development of manufacturers: competitors, governmental regulations, consumer preferences and changes in innovative technologies for implementing SSCM.
- Focus on measures of SSCM: economic and environmental performances.
- Focus on the discussion about the environmental dimension: Green Supply Chain Management (GSCM).
- Define the necessary activities to implement green development in the entire supply chain including green supplier selection, green materials, green transportation and reverse logistics.

- Define the current situation of SSCM in the automotive industry especially of leading car makers in Germany and compare it to Asian OEMs from Japan and China using case studies.
- Analyse the implementation of GSCM in the German automotive industry including challenges, barriers, successes and benefits.
- Estimate the impacts of the changing customers' behaviours on buying battery electric cars instead of internal combustion engine vehicles and customer's expectations on implementing SSCM for the automotive industry's future business by using a survey given to vehicle users.
- Recommend future changing implementations of SSCM for the German automotive industry.

## References

Accenture. (2019). *More than half of consumers would pay more for sustainable products designed to be reused or recycled, Accenture survey finds*. https://newsroom.accenture.com/news/more-than-half-of-consumers-would-pay-more-for-sustainable-products-designed-to-be-reused-or-recycled-accenture-survey-finds.htm

Adams, W. J. (1981). *The automobile industry: In the structure of European industry*. Springer.

Bastin, Z., Bhattacharya, S., & Kumar, A. (2019). *International electric-vehicle consumer survey*. https://www.alixpartners.com/media/13453/ap-electric-vehicle-consumer-study-2019.pdf

Elghali, L., McColl-Grubb, V., Schiavi, I., & Griffiths, P. (2004). *Sustainable resource use in the motor industry: A mass balance approach. Viridis report: VR6*. TRL Limited.

Emmett, S., & Sood, V. (2010). *Green supply chains: An action manifesto*. John Wiley & Sons Inc..

Ernst, D., & Sailer, U. (Eds.). (2015). *Sustainable business management* (1st ed.). UVK.

European Council. (2019). *CO2 emission standards for cars and vans: Council confirms agreement on stricter limits*. https://www.consilium.europa.eu/en/press/press-releases/2019/01/16/co2-emission-standards-for-cars-and-vans-council-confirms-agreement-on-stricter-limits/

Gifford, D. J. (1997). The value of going green. *Harvard Business Review, 75*(5), 11–12.

Hu, A. H., & Hsu, C.-W. (2010). Critical factors for implementing green supply chain management practice. *Management Research Review, 33*(6), 586–608. https://doi.org/10.1108/01409171011050208

Martins, A. (2019). Most consumers want sustainable products and packaging. *Businessnewsdaily.com.* https://www.alixpartners.com/media/13453/ap-electric-vehicle-consumer-study-2019.pdf

Masoumi, S., Kazemi, N., & Abdul-Rashid, S. H. (2019). Sustainable supply chain management in the automotive industry: A process-oriented review. *Sustainability, 11*(14), 3945. https://doi.org/10.3390/su11143945

McCrea, B. (2019). *Driving sustainability in automotive supply chain.* https://www.sourcetoday.com/supply-chain/article/21867380/driving-sustainability-in-automotive-supply-chain

Morana, J. (2013). *Sustainable supply chain management. Automation - Control and industrial engineering series.* ISTE. http://swbplus.bsz-bw.de/bsz381991083cov.htm

PricewaterhouseCoopers. (2007). *The automotive industry and climate change: Framework and dynamics of the CO2 (r)evolution.* http://pwc.com/th/en/automotive/assets/co2.pdf

Rajeev, A., Pati, R. K., Padhi, S. S., & Govindan, K. (2017). Evolution of sustainability in supply chain management: A literature review. *Journal of Cleaner Production, 162,* 299–314. https://doi.org/10.1016/j.jclepro.2017.05.026

Siegfried, P. (2015a). *Business ethics, sustainability and CSR*: Volume 1 - ISBN: 978-3-86924-965-0. AVM Akademische Verlagsgemeinschaft.

Siegfried, P. (2015b). *Business ethics, sustainability and CSR*: Volume 2 - ISBN: 978-3-86924-966-7. AVM Akademische Verlagsgemeinschaft.

Siegfried, P. (2020). *Lebensmittelhandel-business cases Arbeitsfragen and Lösungen*, ISBN: 978-3-75197-990-0. BoD Book on Demand.

Siegfried, P. (2021a). *Enterprise management business*, ISBN: 9783753459011. BoD Book on Demand.

Siegfried, P. (2021b). *Enterprise management automobile industry business cases*, ISBN: 9783753444871. BoD Book on Demand.

WWF. (2019). *What are climate change and global warming?* https://www.wwf.org.uk/climate-change-and-global-warming

# 2

# Traditional Supply Chain Management

## 2.1 Definition of Traditional Supply Chain Management

In the 1980s, the term "Supply Chain" started to appear when many firms realised the benefits from the collaboration and relationships between supplier groups and entities within and beyond their own operations (Díaz, 2006). Instead of supplying sources on their own, companies search for specialised suppliers who can provide lower costs and more qualified materials (Lummus & Vokurka, 1999). The management of networks of suppliers is a tactic for organisations to optimise their overall performance. It leads to win–win situations for suppliers and companies. Another reason is the increase of national and international competition. Customers have more choices to satisfy their demands by using different sources from multiple rivals. Therefore, it is important to manipulate distribution channel networks and inventorying volumes to get maximum customer accessibility with minimum costs. The third reason why the supply chain is becoming more interesting is that organisations realised that optimal performance for the whole of the company may cause more benefits than maximising performances of individual departments or one function. Although it is possible to acquire lower prices on material in purchasing, the inefficiencies in the production can cause higher costs for the business. Consequently, the overview across the entire supply chain is essential to estimate the right decision-making for the company (Lummus & Vokurka, 1999).

Due to benefits and effectiveness from managing the supply chain, the popularity of research on "supply chain" has been rising (Díaz, 2006). According to the APICS Dictionary, supply chain represents the flow of raw material provided by suppliers and processed into finished products or services (Cox et al., 1995; Siegfried, 2013). Quinn (1997) defines that the supply chain consists of activities in different departments like procurement, production, inventory, transportation, warehousing and sales to create end-products from initial materials.

Based on basic theories about the supply chain, supply chain management (SCM) is consequently defined by numeral researchers. Basically, SCM coordinates and manages a complex network of processes in the supply chain involved in supplying products or services to end-users in the most possible efficient and cost-effective manner (Storey et al., 2006). It is a corporate activity which aims at delivering the maximum customer satisfaction in terms of quality and price. Traditionally, SCM focused on maximising economies of scales by using particular practices to meet with stakeholder's desires while the social and environmental concerns were obliterated (Xia & Tang, 2011). Stevens (1989) simplified SCM as an integration of activities like planning, arranging and controlling the flow of materials from suppliers and transforming them into end-products for customers. Bowersox and Closs (1996) added information streams into organisations' supply chain to stimulate the efficiency. A study of Lummus and Vokurka (1999) linked SCM to internal and external partners who cause benefits and harms on companies by their actions. Contributions from all stakeholders like employers, employees, suppliers, customers, partners, competitors and governments ensure that the supply chain processes are being seamless and effective (Lummus & Vokurka, 1999). Figure 2.1 describes the main activities in SCM.

SCM consists of main activities in purchasing, production and distribution (Jammernegg et al., 2009). The purchasing department refers to seeking for materials and resources which are needed for the subsequent

## 2  Traditional Supply Chain Management

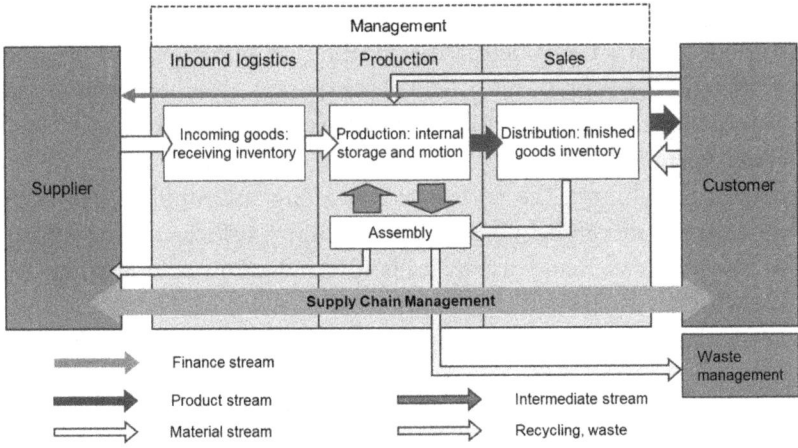

**Fig. 2.1** The activities of supply chain management (own illustration based on Jammernegg et al., 2009)

production steps. Achieving components with the best possible prices from suppliers is aimed for to maximise profitability. On top of that, the purchasing department makes sure of providing sufficient resources and well stocks for production. The production includes converting processes which turn raw material inputs into finished goods or services outputs, which meet the demands and satisfy end-users' requirements. Lower costs and effective management in the production can result in higher levels of efficiency in the supply chain. The distribution transports the finished products to the retail stores, dealers and customers (Lambert et al., 1998; Siegfried, 2021).

These main activities in purchasing, production and distribution connect directly and indirectly with stakeholders in the supply chain which is necessary to establish good relationships between them and leads to win–win situations for all players from suppliers and focal firms to the customers (Bratić, 2011; Kallina & Siegfried, 2021). The implementation of effective supply chain management reinforces competitive advantages for the companies in their industries. In the following the most important supply chain management benefits are listed:

- Better control: When the whole process of business is defined, the company can simply detect the location of materials and products flowing in the supply chain. The delivery time, numbers of offers and purchasing conditions can be controlled quickly and easily (Lorecentral, 2018).
- More profitability: The more controls along the supply chain occur, the more waste can be reduced. The inventory systems will be adjusted to customers' demand which leads to a reduction of operating costs (AIMS UK, 2020).
- Less delays in processes: Managing the supply chain effectively boosts tight cooperation and transparent communication between companies, suppliers and entities that mitigate delays in delivering and production (AIMS UK, 2020).
- Increase in efficiency and competitiveness: When the company integrates supply chain management systems, it will be able to adjust to changing customers' demands and to fluctuating economies. Thenceforth, it strengthens the competitive advantages through eliminating wastes and improves products and services based on customer's needs and values (LEAN, 2020).

## 2.2 Traditional Supply Chain Management in the Automotive Industry

The automotive sector is considered as one of the most important industries in the world and contributes to the economic growth of many countries. The total production has reached 92 million vehicles in 2019. The industry revenues reached 5.35 trillion US dollars in 2017 (Statista, 2020). Counted as one of the biggest industries, the automotive sector has generated employment opportunities for a high number of people globally.

Since the 2000s, the global automotive industry has expanded significantly and transited from domestic to integrated global markets. The manufacturing of most of the input components and materials, which are used for producing end-vehicles, is outsourced to emerging economies like China or India. This trend leads to the establishment of an overly complex network of global suppliers in the automotive industry (Siegfried, 2014). An automotive player can connect with many multi-level suppliers (Arnold, 1997). For example, original equipment manufacturers (OEMs) like Toyota, Honda or BMW have multi-tier suppliers: tier-1 suppliers like Continental, Bosch and Denso and some more specialised tier-2 and tier-3 companies such as ElringKlinger and BorgWarner (Schwarz, 2008).

Making a car is a complex process consisting of a variety of activities like material extraction, component processing, vehicle assembling up to the transport to the final customers. A vehicle can contain more than 20.000 components from many different suppliers (Kapadia, 2018). Globalisation results in higher complexities in supply chain networks of the automotive industry. There are more and more connections between manufacturers, third parties and suppliers from different countries in the world. Since 1985, the contribution of suppliers into a car's production has increased from 56% to 82% (Kallstrom, 2019). The development of SCM showed trends like "Just in time" production in the 1980s and outsourcing as well as global suppliers' collaborations in the 1990s. As measures to increase the cost efficiency, automakers have focused more on assembling pre-produced parts from diverse tier suppliers instead of manufacturing them by themselves (Díaz, 2006). Tier-1 suppliers are most vital suppliers who provide large assembled components directly to the OEM. Tier-2 suppliers supply parts to tier-1 suppliers. The same structure is applied also for tier-3 suppliers. Raw suppliers contribute raw materials to OEMs. Automobiles will be assembled by using these components and raw material. 3PLs are third-party logistics providers who will take over distribution tasks to car sellers or dealers. The structure is shown up in Fig. 2.2.

However, along with the record of notable growth, the automotive industry is still facing many challenges. Today, most cars have engines which burn up fossil fuels during their operation. Consuming big amounts of these fuels leads to hazardous environmental impacts by exhausting harmful substances into the air. The most mentionable of these emissions are carbon dioxide, carbon monoxide and nitrogen oxides. In particular, carbon dioxide causes global warming (Nunes & Bennett, 2010). Additionally, making a car is a complex process

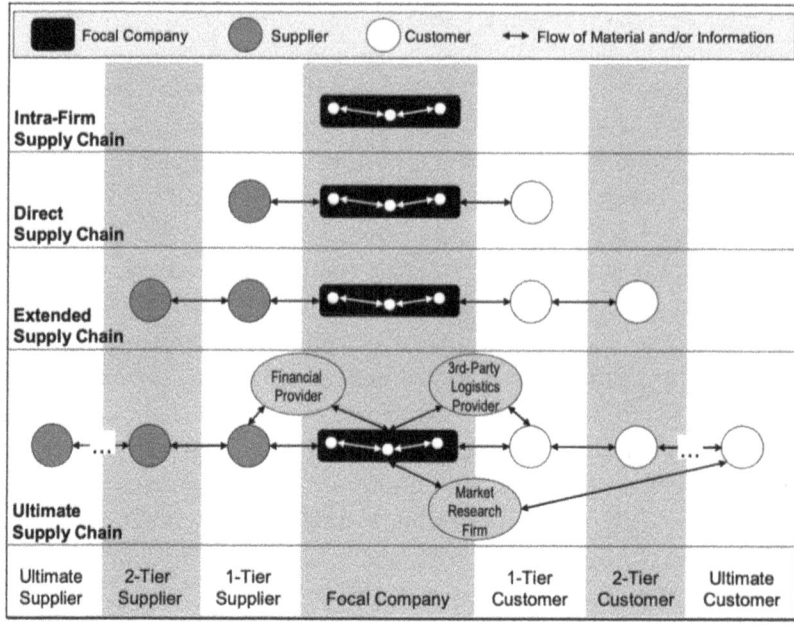

**Fig. 2.2** Levels of supply chain complexity (Díaz, 2006)

consisting of a variety of activities like mining, metal extraction, components production and transportation which all create high carbon footprints. According to a Greenpeace report, all activities of the global automotive industry together are responsible for 9% of global $CO_2$ emissions (Greenpeace, 2019). Therefore, the consumer expectations have shifted towards a demand on eco-friendly vehicles and technological innovations like building up low carbon dioxide emission factories.

At the same time, OEMs are confronted with a high pressure from tightened governmental regulations concerning the reduction of energy consumption, carbon dioxide emissions and safety increase (Rodrigues Vaz et al., 2017). Since the 2010s, stricter governmental regulations, due to the increase of negative effects of the industry to climate change, were showing up and consumers have changed their buying decision towards eco-friendly products (Hunke & Prause, 2014). These transformations have challenged the auto manufacturers to invest in new power-train technologies which offer a better fuel efficiency. In the previous years, to catch up with changes, many OEMs have increased the budgets of their R&D departments to develop e-mobility solutions which use electrical/hybrid power-trains including batteries. Also, they are investing in lightweight and aerodynamic drag-reducing technologies (McKinsey, 2020). Driving sustainability started to become the core strategy to survive and to ensure long-term benefits in the automotive industry.

## References

AIMS UK. (2020). *Key 7 advantages and benefits of supply chain management*. https://aims.education/advantages-and-benefits-of-supply-chain-management/

Arnold, U. (1997). *Beschaffungsmanagement* (2., überarb. und erw. Aufl.). Sammlung Poeschel: 139 Ed. 2. Schäffer-Poeschel.

Bowersox, D. J., & Closs, D. (1996). *Logistical management: The integrated supply chain*. McGraw-Hill.

Bratić, D. (2011). Achieving a competitive advantage by SCM. *IBIMA Business Review Journal, 2011*, 1–13. https://doi.org/10.5171/2011.957583

Cox, J. F., Blakstone, J. H., & Spencer, M. S. (1995). *APICS dictionary* (8th ed.). APICS-The Educational Society for Resource Management.

Díaz, L. M. (2006). *Evaluation of cooperative planning in supply chains: An empirical approach of the European automotive industry*. Deutscher Universitäts-Verlag.

Greenpeace. (2019). *Car industry's 2018 carbon footprint exceeds EU greenhouse gas emissions – Greenpeace*. https://www.greenpeace.org/international/press-release/24131/car-industrys-2018-carbon-footprint-exceeds-eu-greenhouse-gas-emissions-greenpeace/

Hunke, K., & Prause, G. (2014). Sustainable supply chain management in German automotive industry: Experiences and success factors. *Journal of Security and Sustainability Issues, 3*(3), 15–22. https://doi.org/10.9770/jssi.2014.3.3(2)

Jammernegg, W., Kummer, S., & Grün, O. (2009). *Grundzüge der Beschaffung, Produktion und Logistik: Das Übungsbuch*. Wirtschaft. Pearson Studium.

Kallina, D., & Siegfried, P. (2021). Optimization of supply chain network using genetic algorithms based on bill of materials. *The International Journal of Engineering & Science, 10*, 37–47. https://doi.org/10.9790/1813-1007013747

Kallstrom, H. (2019). *Suppliers' power is increasing in the automobile industry.* https://marketrealist.com/2015/02/suppliers-power-increasing-automobile-industry/

Kapadia, S. (2018). *Moving parts: How the automotive industry is transforming.* https://www.supplychaindive.com/news/moving-parts-how-the-automotive-industry-is-transforming/516459/

Lambert, D. M., Cooper, M. C., & Pagh, J. D. (1998). Supply chain management: Implementation issues and research opportunities. *The International Journal of Logistics Management, 9*(2), 1–20. https://doi.org/10.1108/09574099810805807

LEAN. (2020). *What is lean?* https://www.lean.org/whatslean/

Lorecentral. (2018). *Advantages and disadvantages of supply chain management.* https://www.lorecentral.org/2018/12/advantages-and-disadvantages-of-supply-chain-management.html

Lummus, R. R., & Vokurka, R. J. (1999). Defining supply chain management: A historical perspective and practical guidelines. *Industrial Management & Data Systems, 99*(1), 11–17. https://doi.org/10.1108/02635579910243851

McKinsey. (2020). *The road to 2020 and beyond: Whats driving the global automotive industry.* https://www.mckinsey.com/industries/automotive-and-assembly/our-insights/the-road-to-2020-and-beyond-whats-driving-the-global-automotive-industry

Nunes, B., & Bennett, D. (2010). Green operations initiatives in the automotive industry. *Benchmarking: An International Journal, 17*(3), 396–420. https://doi.org/10.1108/14635771011049362

Quinn, F. J. (1997). What's the buzz? *Logistics Management, 36*(2), 43–71.

Rodrigues Vaz, C., Shoeninger Rauen, T., & Rojas Lezana, Á. (2017). Sustainability and innovation in the automotive sector: A structured content analysis. *Sustainability, 9*(6), 880. https://doi.org/10.3390/su9060880

Schwarz, M. (2008). *Trends in the automotive industry implications on supply chain management.* https://www.cisco.com/c/dam/en_us/about/ac79/docs/wp/ctd/Auto_Trends_WP_FINAL.pdf

Siegfried, P. (2013). The importance of the service sector for the industry. In *Teaching crossroads: 9th IPB Erasmus week* (pp. 13–23). Instituto Politécnico de Braganca, ISBN: 978-972-745-166-1.

Siegfried, P. (2014). Analysis of the service research studies in the German research field. In *Performance measurement and management* (pp. 94–104). Publishing House of Wroclaw University of Economics, ISBN: 978-83-7695-473-8, Band 345. https://doi.org/10.15611/pn.2014.345.09

Siegfried, P. (2021). *Handel 4.0 Die Digitalisierung des Handels - Strategien und Konzepte 1*, ISBN-13: 9783754345030. BoD Book on Demand.

Statista. (2020). *Global automotive industry revenue between 2017 and 2030.* https://www.statista.com/statistics/574151/global-automotive-industry-revenue/

Stevens, G. C. (1989). Integration of the supply chain. *International Journal of Physical Distribution and Logistics Management, 19*(8), 3–8.

Storey, J., Emberson, C., Godsell, J., & Harrison, A. (2006). Supply chain management: Theory, practice and future challenges. *International Journal of Operations & Production Management, 26*(7), 754–774. https://doi.org/10.1108/01443570610672220

Xia, Y., & Tang, L.-P. (2011). Sustainability in supply chain management: Suggestions for the auto industry. *Management Decision, 49*(4), 495–512. https://doi.org/10.1108/00251741111126459

# 3

# Sustainable Supply Chain Management

## 3.1 Sustainable Development

### 3.1.1 Definition and Advantages of Sustainability

Human society has developed constantly with technological advances and global integration. Over the last decades, the world's GDP has grown in average over 3% per year, reaching to double its size by 2037 and tripling in 2050 (PricewaterhouseCoopers, 2020). The escalation of industrialisation to achieve world economic growth has caused environmental problems like global warming, increasing greenhouse gas emissions, air and water pollution, growing volumes of waste, desertification and chemical pollution. Industrial processes have played a major role in global environmental degradation (Ahuti, 2015).

By increasing awareness on negative impacts from extremely polluting industrial activities, nature protection and carbon dioxide reduction trends have been emergent as business strategy directions for all types of industries over the world. Therefore, manufacturing companies strive to reduce harmful effects on the environment through measures like adopting eco-friendly product designs and operating environmental practices.

Based on the occurrence of many ideas in environmental sciences, in 1987, the Brundtland Commission of United Nations published its report with the topic "Our Common Future" in which the term "sustainable development" was first defined as a "development that meets the needs of the present without compromising the ability of future generations to meet their own needs" (McGill, 2020). Recently, the concepts of sustainability and sustainable development have spread and appeared in various research areas and literature sources.

Sustainability means the ability to maintain a process or a state at an enduring certain level. It is important that humans use natural resources only with environmental awareness (Leung, 2020). In a broader meaning, sustainability is not only just focusing on environmentalism but also concerned with other factors like social and economic dimensions (McGill, 2020). Sustainable development seeks for long-term economic benefits without creating negative impacts on the environment as well as society and culture. It seemed that the combination and harmonisation of these three dimensions ensures the long-lasting prosperity growth of the world (McGill, 2020).

Besides diminishing overall negative impacts on the environment sustainable practices generate other benefits for organisations, especially for the manufacturing industry which uses large quantities of materials and resources and emits huge amounts of pollutants and $CO_2$ gases:

- Cost reduction: The scarcity of non-renewable resources pushes increasing prices on purchasing materials. Hence, sustainable practices give more chances to companies to save costs through using renewable materials in manufacturing and saving energy and reducing waste by using technological solutions (Leung, 2020).
- Brand images: Consumers have become more concerned with buying sustainable products. In order to gain bigger target audiences, companies operate their strategies sustainably and use sustainable manufacturing techniques. This builds up their reputation on environmental awareness and leads to higher sales and revenues (Leung, 2020).
- Competitive advantages: Establishing the image of an environmentally conscious manufacturer helps the company distinguishing with competitors who are just focusing on profits (Brown, 2017).

- Promote innovation: Committing to sustainable development, companies face with challenges in innovation improvements. For example, novel technologies for $CO_2$ emission reduction in the production process need to be developed (Brown, 2017).

### 3.1.2 Sustainability Framework: Triple Bottom Line (TBL)

The measures of sustainable development consist of three dimensions: economic, social and environmental. These core dimensions have been determined in the Triple Bottom Line (TBL) concept which refers to 3Ps: people, plant and profit (Elkington, 1998). Based on the TBL framework, the successful performance of an organisation is assessed not only by financial benefits, but also by its environmental consciousness and ethical values (Gimenez et al., 2012). Many studies define TBL as a method to help firms to maintain their survival ability in the long run. Organisations could lose customer interests if they only focus on economic development and neglect positive activities on environmental and social issues (Carter & Easton, 2011).

#### 3.1.2.1 Economic Dimension

To obtain successful sustainable performance in the long term, operational activities need to be profitable. The economic aspect describes the monetary flow in the company and is measured by expenditures, employment incomes, costs, revenues, etc. (Slaper & Hall, 2011). From the economic point of view, sustainable development refers to activities that create financial values contributing to the growth of firms in particular and growth of entire economic system in general (Arowoshegbe & Emmanuel, 2016). Based on the TBL framework, the economic pillar cooperates harmoniously with environmental and social considerations. The economic-social domain accounts for actions related to profitable and ethical values like fair trade, business ethics and workers' rights. Furthermore, the economic-environmental domain takes the decrease of harmful effects on the natural environment into account through following up with economic achievements like reducing costs by using energies efficiently (Carter & Rogers, 2008).

#### 3.1.2.2 Environmental Dimension

The environmental dimension focuses on reducing negative impacts on the environment by cutting down $CO_2$ emissions, utilisation of natural resources, recycling and waste disposal (Arowoshegbe & Emmanuel, 2016). A study of Vachon and Klassen (2008) shows that in order to attain ecological benefits, manufacturers need to apply environmental concepts in collaboration with suppliers and develop eco-friendly innovative technologies. Furthermore, the governmental legislation on nature protection is triggering environmental performances of companies in SCM.

#### 3.1.2.3 Social Dimension

The social line in TBL refers to beneficial values which the firms bring to the society. Examples for social practices include ensuring fair wages, health care coverages, improving working conditions and accident prevention. Training knowledge and employees' skills play important roles for future human development. Moreover, charities and donation events are ways to build up the company's prestige to gain the customers' trust (Goel, 2010).

### 3.1.3 Regulations for Sustainable Development

Dealing with growing threats of global warming and its negative impacts on the environment, the United Nations Framework Convention on Climate Change (UNFCCC) entered in 1994 with 197 countries taking part in it. The objectives of the UNFCCC are to "stabilize greenhouse gas concentrations in the atmosphere at a level that would prevent dangerous anthropogenic interference with the climate system" (UNFCCC, 1992).

In 1997, the Kyoto Protocol was signed in Kyoto, Japan, which was linked to the UNFCC and aimed at reducing carbon emissions and greenhouses gases in the atmosphere. There are 175 states admitting to the Kyoto Protocol and mandating to degrade the amount of their $CO_2$ emissions. These industrial nations were assigned maximum carbon emission levels for a certain time period (UNFCCC, 2008).

As being well-known as a leader in environmental protection activities, the European Union has developed and encouraged various policy tools to ensure sustainable production and consumption patterns. The EU has reinforced environmental legislation and standards that aim to the sustainable development in the long term like preservation of nature's resources, protection of the well-being of people and fulfilment of economic benefits (EUR-Lex, 2020).

In 1992, the EU eco label was created to certificate products which are manufactured environmentally friendly. This "EU Flower" label has been used on more than 77.000 products and has broadened widely in EU member states (European Commission, 2019). To maintain sustainable development in the long term, the EU launched a $CO_2$ emissions trading scheme which encourages companies to develop low-carbon technologies and to obtain effective costs by reducing emission levels. This scheme permits participating companies a certain level of $CO_2$ emissions per year and promotes them on investing in clean, low-carbon technologies to ensure their emission allowance certificates. The purpose of the EU Emissions Trading System (EU ETS) in 2020 is that the emissions will be 21% lower than in 2005, and, in 2030, they should be cut by 43% compared to the 2005 levels (European Commission, 2020).

## 3.2 Definition of Sustainable Supply Chain Management

As mentioned above, the supply chain has become a key success factor for all industries. It brings more profitability and competitive advantages to businesses. Quality and costs have been a focus of supply chain management for a long time. However, over the past years, due to impacts from climate change and from consumer buying behaviour, the traditional supply chain management has shifted towards a sustainable development (Siegfried & Zhang, 2021). Therefore, the definition of sustainable supply chain management (SSCM) has been researched in many studies.

Sustainable development is not just restricted to a green factor but deals with factors like social responsibilities or economic benefits too (McGill, 2020). Adopting the sustainability approach, SSCM can be defined based on the concept of the Triple Bottom Line (TBL) with the integration of environmental, social and economic criteria which is illustrated in Fig. 3.1.

According to a study of Carter and Rogers (2008), SSCM is defined as "the strategic, transparent integration and achievement of an organisation's social, environmental, and economic goals in the systemic coordination of key interorganisational business processes for improving the long-term economic performance of the individual company and its supply chain". It is emphasised that in order to acquire sustainability, companies need to reach equality of optimising monetary benefits, protecting

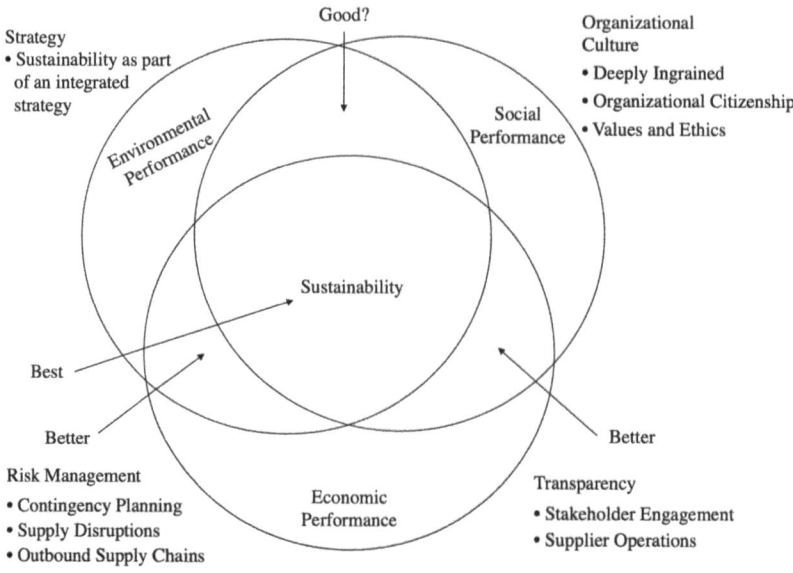

**Fig. 3.1** Sustainable supply chain management (Carter & Rogers, 2008)

natural resources and satisfying social needs which is depicted as "Best" in Fig. 3.1 within the intersection of environmental, social and economic performance. The question mark between social and environmental shows that leaving out the economic goals can harm the survival of companies. Carter and Rogers assert that the social and environmental dimensions of SSCM must be performed with an explicit recognition of the economic objects of the firm (Carter & Rogers, 2008).

The implementation of sustainable development in SCM provides competitive advantages for organisations (Beske et al., 2014). SSCM manages the flow of materials and goods with the purpose of the minimisation of harmful impacts on the environment but still creates economic advantages and contributes to social responsibilities at the same time (Hu & Hsu, 2010). Activities like green materials purchasing, reduction of waste, saving resources, using renewable energies, recycling and disposal commit to the sustainable supply chain network (Seuring & Müller, 2008).

However, equality of the three dimensions is hard to reach. It has been claimed that the environmental and social responses put more challenges on saving costs for companies (Walley & Whitehead, 1994). For instance, in order to reduce natural resources usage, companies must spend money on renewable alternative resources which might endanger their financial status (Rogers et al., 2007). Therefore, it is important for manufacturers to undertake economic, social and environmental initiatives along the supply chain management in bearable, viable and equitable ways (Carter & Rogers, 2008).

Furthermore, according to a study of Pagell and Gobeli (2009), it is concretised that SSCM integrates firms' plans and activities across their supply chain networks with environmental and social objects in order to improve not only companies' sustainability performance but also that of their suppliers and customers. Based on this definition, SSCM extends boundaries of companies and also covers sustainable performances of different stakeholders like suppliers, customers and governments who influence and trigger the implementation of SCM.

Other than crucial economical goals, like profit maximisation, there is also pressure put on the focal companies by stakeholders to adopt sustainable development on their supply chains. Governmental pressures can be local, national and international standards and regulations.

Customers' motivations originate from their buying behaviour shifting towards preferencing eco-friendly products. Environmental technologies of competitors play a significant role for focal company during developing sustainable innovations. These are pushing factors forcing the company to apply sustainable performances. If the focal company performs sustainably, it fosters pulling factors for suppliers to be involved in sustainable development (Rebs et al., 2018). Therefore, in order to reach win–win relationships between economic, environmental and social goals and supply chain's players, the focal company should realise these activities for SSCM practices such as sustainable suppliers training, sustainable risks and pressure management (see Fig. 3.2) (Brandenburg & Rebs, 2015).

## 3.3 Driving Force Factors

The global environmental hazards have led various stakeholder groups to demand more environmentally friendly products and services. Stakeholders are key persons who were pushing factors on manufacturers

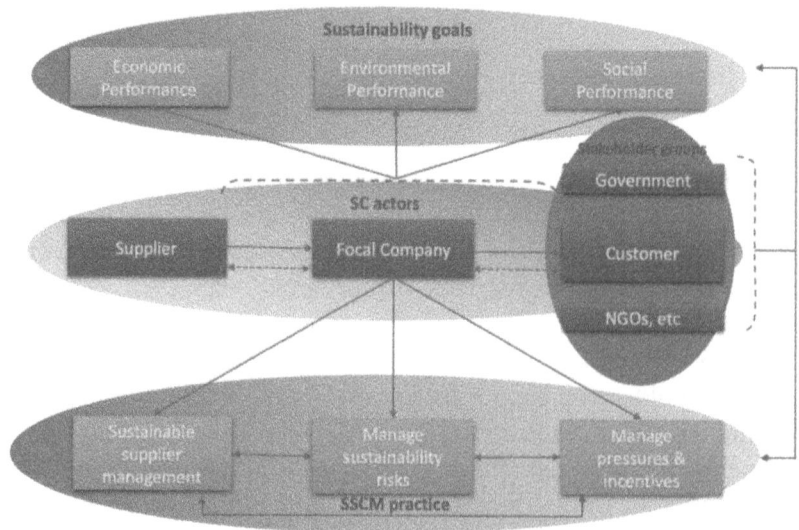

**Fig. 3.2** Supply chain sustainable framework (Brandenburg & Rebs, 2015)

in the supply chain sustainability shift. The expectations of stakeholders put more pressure on firms to adopt sustainable practices along their supply chains (Zhu et al., 2013). Each of these factors who push companies to pursue SSCM is discussed in the following sections.

## 3.3.1 Customers' Behaviour

All activities of SCM lead to the delivery of final products to end customers. Customers are a stakeholder group who directly affect to profitability and the financial performance of the firm (Apte & Sheth, 2017). The more customers can be gained, the more revenues can be reached. Therefore, measuring and enhancing the customer satisfaction can increase customer loyalty and retention (Scott et al., 2018). Unlike the past, the customers are obtaining awareness about the origins of the products they buy, from where they were made, who made them, under which conditions and when they were delivered (Thompson, 2020). Accordingly, the customer is the key figure in the supplier chain and their values and opinions impact on the companies' decision-making. Handfield and other authors asserted likewise that customers' needs and expectations are important for the creation of supply chain initiatives (Handfield et al., 2005; Siegfried, 2017).

Due to negative ecological changes like global warming, scarcity of natural resources and the degradation of the living environment, people have been increasing their awareness on environmental and social issues which lead to a transformation in the consumers' buying behaviour. More and more consumers tend to include environmental performances into their purchasing decisions (Russo et al., 2015). A survey of Global Consumer Confidence showed that more than 81% of global respondents would prefer buying products or services from companies who engage in social and environmental actions (Nielsen, 2018). For instance, in the automotive industry, low-carbon emissions, hybrid fuels and electrification are becoming crucial core buying criteria. Other than eco-friendly cars, recyclable materials and components from the supply chain processes attract attention of green consumers. Because of this change in behaviour, sustainability progress is a new fundamental part of the supply chain of automotive companies if they do not want to deteriorate their reputation (Sarkis et al., 2010).

With the increase of environmental responsibility of customers on buying products, the companies might fear boycotts if they are not responsible with the environment. That could lead to image damage and financial losses (Sarkis et al., 2010). Customer pressure is, therefore, the driving factor contributing to companies' sustainable supply chain adaptation (Hsu et al., 2013). Moreover, Porter (2008) also verifies that customer pressure is also a prior condition forcing government to establish environmental regulations in the first place. By acting together, customer and regulation forces become comprehensive pressures on companies to practise their supply chain with environmental protection initiatives.

### 3.3.2 Governmental Regulations

In the past, companies played an essential role in managing their systems and defining their own regulations, while government roles were put on the side-line (Seipp et al., 2020; Vermeulen et al., 2011). Nowadays, negative impacts of the climate change and resource shortages lead to the imposition of stricter governmental environmental regulations on manufacturers, who are the main resource consumers and polluters (Zhu et al., 2013). Government agencies become power groups who enact regulations to obligate manufacturers in the involvement with environmental responsibilities and force them to include sustainability into their SCM. In exchange, using renewable resources and environmental-friendly technologies generates competitive advantages and an efficient performance for them.

Companies might ignore other stakeholders' demands about environmental responsibilities, but governments leave companies no freedom of choices (Güner & Coskun, 2010). Governmental pressures such as regulatory enforcement are defined as coercive pressures which are considered the most influential pressures towards adoptions of environmental initiatives (Jennings & Zandbergen, 1995). According to Bansal (2005), the failure of compliance with governmental regulations may result in many disadvantages for the companies like image and reputation damaging, loss of operation licences or facing of legal sanctions. Governments may enforce taxes for companies not practising sustainability in their supply chain management (Clemens & Douglas, 2006).

In the last decade, many stringent regional and national regulations have been enacted to support environmental initiatives. For example, the European Union has developed regulations and environmental directives like End-of-Life Vehicles (ELV), Restriction of Hazardous Substances (RoHS), Energy-Using Products (EuP), Waste Electrical and Electronic Equipment (WEEE), etc. (Koh et al., 2012). These legislations have obliged manufacturers to take back or recover resources or products after their usage with the goal to reduce waste disposals. "Remanufacturing" in the production is a process where products are taken apart, repaired and reassembled to be used again. This supports saving energy and reducing wastes (Scott et al., 2018).

Besides coercive pressures, governments also use motivating strategies to encourage focal/local companies in adopting environmental practices. The financial incentives such as tax reduction or subsidies have been offered as a support to companies to invest more voluntarily in sustainable supply chain by many governments (Boström et al., 2015). For example, the German government supports consumers with a subsidy when they purchase hybrid or electric cars. In addition, tax dispensations for 10 years are covered for all electric cars (Fullerton, 2017). In the UK the government announced tax reduction policies for using biofuel in order to promote sustainable growth (Kumar & Joo, 2019). Similarly, the US government offered financial benefits and grants for entrepreneurs who are related to environmentally responsible approaches (Green Business Bureau, 2019).

### 3.3.3 Competitors

Some studies show that competition is another motivation to integrate sustainable strategies into the supply chain progress. Today, the global market has developed widespread challenges with many new entrants which leads to an intensity of rivalry between firms. Traditionally, they compete with others based on aspects like price, quality, promotion or service (Siegfried, 2014). Hence, the differentiation from their competitors is the value to gain competitive advantages for companies (Saeed & Kersten, 2019).

Because of an image boost and higher customer satisfaction, companies who undertake environmental initiatives and gain a competitive edge in the market will pressure other companies to follow and imitate their environmental strategies (Rivera, 2004). Success of competitors who are practising environment protection strategies is determined as pushing factors for the other firms to imitate environmental practices. Following competitors' actions is regarded a trigger for sustainable opportunities that enhance competitive abilities for firms (Zhu et al., 2010).

### 3.3.4 Innovative Technology Development in Sustainable Supply Chain

Technological innovation is seen as another factor leading to sustainable development. Environmental regulations and changed consumers' consciousness impulse new eco-friendly innovative breakthroughs (Rodrigues Vaz et al., 2017).

Next to economic goals, innovative technologies are also seen as leading paths to optimal sustainable development in the entire supply chain processes (Pereira de Carvalho et al., 2012). With the increase of the technological evolution, it is important to gain business success by integrating technologies successfully in the entire business operation and supply chain and enhance the competitive advantages for the company (Artsiomchyk & Zhivitskaya, 2015).

Suppliers are important partners to manufacturers, as their works impact directly and extensively on the cost, quality, technology and time-to-market of new products (Handfield et al., 1999). In the early stage of the supply chain, by selecting suppliers who carry out optimal technologies, it will help manufacturers to improve the resulting end-product. The instructions and knowledge of suppliers inspire greater awareness for companies in adoption of new technologies in their manufacturing (Jantan et al., 2006). Therefore, in order to get achievements on environmental solutions, manufacturers need to find suppliers who also operate with sustainable technologies.

According to Kemp and Arundel (1998), there are technological solutions related to environmental protection, for example: end-of-pipe technologies, clean-up technologies, waste management technologies and recycling technologies. By using environmental technologies in the manufacturing processes, it helps to reduce the consumption of energy and resources, to prevent polluting emissions and to recycle wastes (Qudrat-Ullah, 2018).

To fulfil customers' environmental concerns, many companies put more effort in researching and developing new green products. Innovative environmental technologies applied in supply chain processes contribute to the companies' abilities of producing eco-friendly products which meet the ecological targets and requirements (Qudrat-Ullah, 2018). The production of green products brings competitive advantages for businesses if they differentiate from others in the market. Furthermore, it contributes to an overall green company image towards the customers (Qudrat-Ullah, 2018).

## 3.4 Performance Measures for Sustainable Supply Chain Management

Based on the triple bottom line concept, SSCM performances intersect the three dimensions: environmental, social and economic. Organisation performances which are in conjunction to TBL are assessed as successful not only by a traditional financial-based point of view, but also by the environmental and ethical consciousness (Gimenez et al., 2012). The integration of economic, environmental and social performances aims to decrease harmful ecological impacts and increases positive influences on the society while achieving long-term economic and competitive advantages (Saeed & Kersten, 2019; Siegfried, 2015).

In relation with the TBL concept and in order to ensure a successful SSCM, it is required to effectively measure all three key dimensions (Hervani et al., 2005). In many literature sources it is suggested that companies' improvements in sustainable supply chain performance (SSCP) can sharpen their competitiveness and financial and operational performance. SSCP is a company's capacity to reduce the use of materials, energy or water and to find more eco-efficient solutions by improving the supply chain management (Figge et al., 2002). Therefore, following this approach, this research will examine SSCM intensively with economic and environmental measures.

## 3.4.1 Economic Performance

Economic performance is measured by indicators which impact directly to the financial status of a firm. Generally, these indicators are noted as profits, market shares, sale revenues, growth, etc. Financial performance is defined as long-term objectives leading to success for the operation (Kaplan & Norton, 1996). According to Schaltegger and Synnestvedt (2002), a company that just focuses on environmental development but not on financial advantages will later be evaporated from the market along with its environmental beneficial activities. Therefore, economic indicators are important for a company's survival. Furthermore, satisfying stakeholders' requirements on financial goals will improve long-term economic performance of the corporation. By using their financial powers for adapting to changes in external demands such as new technologies or product developments the company can also enhance its efficiency and its competitive advantages (Freeman & Evan, 1990). However, the most effective approach to assess financial performance in the adoption of SSCM is the cost reduction that is associated with energy consumption, purchased materials, waste and disposal treatment (Green et al., 2012). Transforming waste into resources is a method to help companies in improving their financial performance. Moreover, including the participation of suppliers into decision-making processes of environmental innovations allows companies to reduce purchased material costs (Ortas et al., 2014).

To measure successful sustainable performance, it is necessary for companies to combine their financial performance with environmental initiatives. According to Rao and Holt (2005), by undertaking environmental responsibilities across the supply chain, eco-friendly customers' needs will be satisfied, the corporate reputation will be built up which can turn into economic benefits like higher sales and profitability. Despite having possible positive financial outcomes, many empirical papers assume that the adoption of environmental initiatives may also cause negative effects on companies' financial status in the short term (Hahn & Figge, 2011). For example, to carry out environmental actions, companies need to invest in eco-friendly innovative technologies to produce green products or to reduce waste disposals. This can generate higher costs for the company.

### 3.4.2 Environmental Performance

The expectations of customers using eco-friendly products, the associated compliance with governmental regulations on ecological protection and the competitive pressure from competitors are reasons why manufacturing companies pursue the adoption of eco-designs into the SCM context. In this regard, enhancing environmental performance is the measure leading to improvements in a sustainability-based supply chain (Zhu & Sarkis, 2007). The environmental performance is related to reduction of negative impacts on the environment like $CO_2$ emissions, waste-, energy-, water- and hazardous material consumption (Esty & Winston, 2009). Although environmental outcomes are non-economic performances because they do not directly create financial benefits, they can still influence the economic success of the company, for example, by attracting more customers with their green image (Zhu & Sarkis, 2007). Many literature sources describe the tight relationships between environmental and economic performance in the implementation of SSCM (Rao & Holt, 2005).

By understanding that the environmental performance is regarded as a crucial issue for sustainable development, many organisations have added environmental management programmes along the supply chain which are managed by an Environmental Management System (EMS). As stated in ISO (1996), EMS is defined as "an integral part of an overall management system that includes organisational structure, planning, activities, responsibilities, procedures, processes and resources for developing, implementing, achieving, reviewing and maintaining environmental policy". EMS is an aiding tool for companies to approach environmental protection measures and to estimate impacts of companies' activities on the environment. The goals of EMS are to increase compliance of companies on environmental policies and to eliminate negative impacts on the environment through reducing waste, prevention pollution and recycling (Sroufe, 2003). There are various standards in EMS such as ISO 14000, which provides mechanism to manage and improve environmental performances. Additionally, there is EMAS (Eco-Management and Audit Scheme) introduced by the European Union Council which requires all member states mandatorily to recognise environmental actions.

## 3.5 Causes for Implementing Sustainable Supply Chain Management in the Automotive Industry

As shown in Sect. 3.1.2, sustainable development is defined by the three dimensions: economic, social and environmental. Sustainable supply chain management is based on the principle of supply chain management with an extra add-on of green impacts, which means environmentally friendly and efficient aspects. SSCM aims at providing the logistic aspects of the production process in the company in the most efficient way. This includes the approach on how ecological aspects can be considered in the whole business processes in the most effectively way (Hunke & Prause, 2014).

In the automotive industry applying sustainability not only covers the profitability of OEMs, but it also defines new key performance indicators for vendors and suppliers in the supply chain network. There are more than 25.000 suppliers from 100 countries joining the "Drive Sustainability Platform" that orientates on activities, requirements and common projects towards sustainability for supply chains (Schwarzkopf & Dorwald, 2019). The supply chain processes of a vehicle, from components selection over transportation and manufacturing to end customer delivery, all effect the overexploitation of natural resources and the high level of $CO_2$ emissions. To eliminate these harmful effects, most automotive leaders need to practise sustainable activities in their supply chain. It can be assumed that the involvement of green aspects in the supply chain of a company also initiates changes in the supply chain itself. The successful

SSCM implementation embraces social and environmental matters, next to the economic benefits (Carter & Easton, 2011).

As defined in Sect. 3.3 there are driving factors leading companies to practising sustainable supply chain management or green supply chain management: competition, customers' behaviour and governmental regulations. Based on these factors the following chapters will examine causes for implementing SSCM especially in the automotive industry.

### 3.5.1 Growing Competitive Markets

Automotive is a complex industry with intensive competitive pressures from many giant global rivals from different continents. In Europe, there are pioneers in car manufacturing like Daimler, BMW or VW. In the Asian markets, Toyota and Hyundai are leaders in producing a variety of car models targeting middle income population groups. Ford, GM and Fiat Chrysler are well-known as the big three in the USA. Especially the American electric-automobile manufacturer Tesla, which was recently founded in 2003, has overtaken German automotive makers by becoming the world's number two automaker by market capitalisation (Richter, 2020). Figure 3.3 describes the capitalisation of different automotive brands.

| Market Capitalization | |
|---|---|
| | Billion $ |
| Toyota | $233.9 |
| Tesla | $102.0 |
| Volkswagen | $97.6 |
| Daimler | $53.6 |
| Honda | $49.8 |
| BMW | $49.5 |
| GM | $49.0 |
| Ford | $36.0 |
| FCA | $20.9 |

**Fig. 3.3** Market capitalisation of automotive brands (Richter, 2020)

According to forecasts of the International Energy Agency (IEA), the number of cars will not grow anymore in the industrialised countries. The BRIC (Brazil, Russia, India, China) states will soon overtake Europe and the USA here. Currently, China is becoming the largest market for commercial and passenger vehicles (Richter, 2020). The pressure from environmental protection regulations has caused Chinese automotive makers to consider the implementation of SSCM. Many automotive OEMs in China have been aware of the scarcity of natural resources and the increasing greenhouse gas emissions. There are reasons which push them to develop and produce vehicles with new engine technologies. The new engine vehicle (NEV) industry emerged as a strategic industry with tasks related to reducing environmental pollution, reducing fossil energy consumption and the support of the ongoing production of vehicles within the automotive industry (Wu et al., 2018). NEVs are defined as vehicles using unconventional fuel power sources or new power devices: These are, for example: battery electric vehicles (BEVs), plug-in hybrid electric vehicles (PHEVs) and fuel cell vehicles (FCVs). At the end of 2017, China has recorded a parc of 1.53 million NEVs which equals 50% of the global total ownerships. The Chinese automobile enterprises, such as BYD, BAIC, Geely and CHANA, are pioneers on implementing new engine technologies in their cars. For example, BYD has been number one in the world's NEV sales for the last three years. Both BAIC and CHANA have planned to stop selling traditional fuel automobiles before 2025 (Wu et al., 2018).

Not only Chinese, but also Indian automobile enterprises have embraced green processes in their production in the recent years. Tata Motors has set its mission to produce carbon neutral vehicles. In order to reduce climate change effects in the manufacturing, Tata has developed green dealership concepts which boost up green awareness and promote practising optimal environmental management in their supply chain systems (Telang, 2013). Japanese automakers have entered green car markets very early. Leaders like Toyota and Honda have been practising power-train development for green energy vehicles. The Toyota Prius, the first hybrid electric vehicle to go into serial production, was launched in the Japanese market in 1997 (Jürgens & Meißner, 2005).

Therefore, in order to maintain competitive advantages and leading positions in the automotive sector, German companies need to develop innovative, environmentally friendly mobility concepts and green supply chain systems to ensure a future sustainable survival.

## 3.5.2 Green Consumers

As mentioned before, green awareness among consumers is the most important force for companies to practise green supply chain management. Customers' demands influence directly to sales and revenues of enterprises. Many studies have shown that customers have been changing their preferred way of travelling and their future mobility expectations (Deloitte, 2018). In the automotive industry, low-carbon emissions and electrification are becoming crucial core buying criteria. Besides eco-friendly cars, recyclable materials and components from the supply chain processes attract the attention of green consumers. Because of this behaviour change, progress in sustainability is a new fundamental part of the supply chain of automotive companies if they do not want to deteriorate their reputation (Sarkis et al., 2010; Siegfried & Strak, 2021).

Although production costs of green cars are high, there is still an increase in the number of eco-friendly vehicles sales since 2010. According to the study of IEA the global electric car fleet exceeded 7.2 million in 2019 (IEA, 2020). That is 2.1 million more than in the previous year. In Fig. 3.4 the yearly values are shown starting from just under 500,000 units in 2013.

Figure 3.5 shows the yearly deliveries of BEV and PHEV from 2010 to 2019 which also show steady high growth rates and reached 2.2 million deliveries in 2019. In 2019, this growth rate dropped significantly for the first time in years which is caused by a decrease in sales in China and the USA, which are the two largest markets (Virta, 2020).

Looking at the different countries, China remained the world's largest EV market in 2019 which accounted for 3.4 million electric vehicles running on their roads. This is equivalent to nearly 47% of the global EVs. Europe and the USA are relatively standing behind with 1.8 and 1.5 million EVs in use (IEA, 2020).

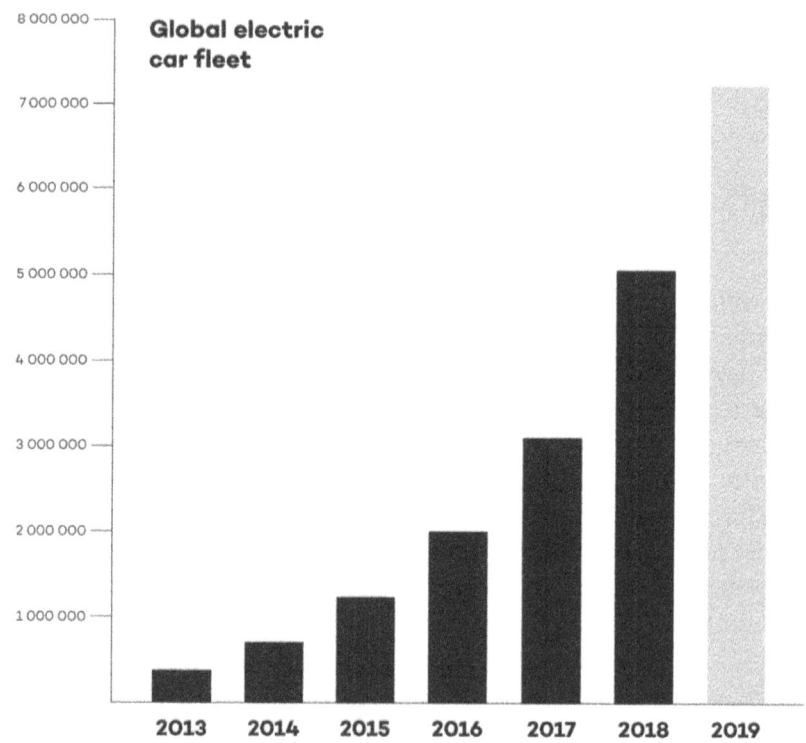

**Fig. 3.4** Global electric car fleet (Virta, 2020)

### 3.5.2.1 Survey about Automotive Customers' View on Sustainable Supply Chain Management

To find out more about the consumers' view on sustainable supply chains in the automotive industry, a customer survey was performed during the creation of this research. It should help to answer the questions whether the consumers show a higher interest in products with green supply chain characteristics, whether they are aware of the origins of the materials, whether they are renewable or recyclable and whether the availability of suppliers with green performances is even taken into consumers' considerations. The results can be found in the appendix. A questionnaire was

3  Sustainable Supply Chain Management    37

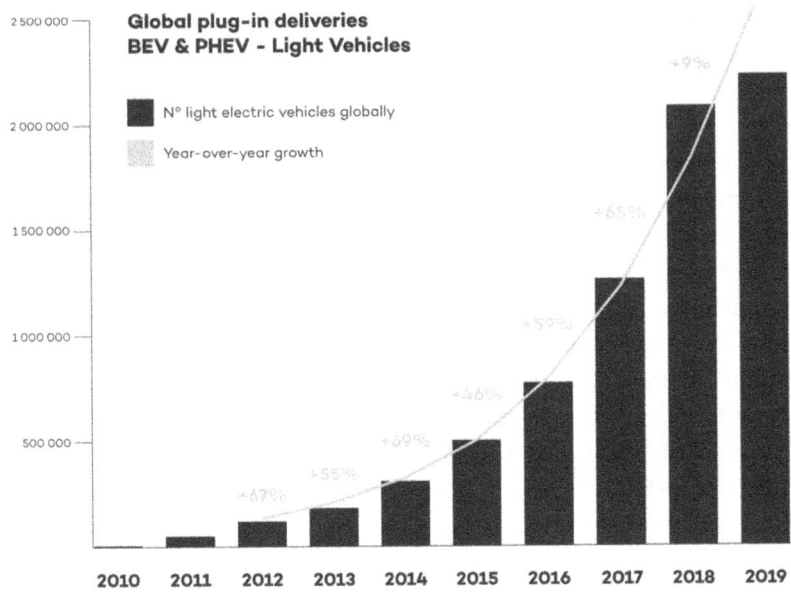

**Fig. 3.5** Global BEV and PHEV deliveries (Virta, 2020)

sent out and collected 101 usable responses. The participants are distributed to both male and female gender and to different age groups.

From the findings of the executed survey, it can be assumed that it is important for the majority of vehicle customers that automotive manufacturers pursue sustainable development strategies. Moreover, they are becoming more aware with the sustainability of the entire lifecycle of vehicles that can reduce negative effects on the environment. Additionally, the interest of customers on SSCM increased within the last years combined with growing intentions on purchasing green vehicles.

Another main finding from the customer survey performed in this Research is that the consumers would even accept higher prices for vehicles with a sustainable supply chain. Especially the customers that also research about the supply chain of their desired vehicles are willing to pay more for these cars if they find out about a sustainable supply chain. These are crucial information to be considered in companies' development plans.

### 3.5.3 Environmental Global Laws

The automotive industry is perceived as a large source leading to greenhouse gas (GHG) emissions from its produced conventional automobiles which are operated by internal combustion engines. It was analysed that the transport sector is responsible for 25% of the total carbon emissions in the EU (Frondel et al., 2011). By understanding the negative effects from car manufacturing, many specific measures have been formulated in both the international and the EU level such as climate policies and regulations for environmental protection. These are factors pushing automotive manufacturers to implement the green supply chain.

One sample of these ecological laws is the packaging and waste directive which was applied in the European Union since 1994 or the German Act on closed-loop resource management in 1996. These regulations provided legal frames which direct companies to reduce waste and to protect natural resources during production processes and the life cycle of the product (Hunke & Prause, 2014). Furthermore, the EU has imposed new standards for $CO_2$ emission limits from vehicles and its components. According to these targets, all new cars sold in the EU should not emit more than 95 g/km of $CO_2$ by the end of 2020. Paying a penalty is the punishment for manufacturers who do not achieve these standards (Kotak & Kotak, 2016). Germany is one of the few EU member states that have been successful in following the newer strict regulations about greenhouse gas emissions by reducing their emissions produced by transport. Compared to 1990, Germany agreed to reduce its greenhouse gas emissions by 21% by 2012 (Friedrich Ebert Stiftung, 2015).

Energy labelling is also an optimal method helping consumers to identify energy-efficiency information of automobiles. Hence, the EU has implemented the vehicle labelling directive 1999/94/EC which provides "information relating to the fuel economy and $CO_2$ emissions of new passenger cars offered for sale or lease in the community in order to enable consumers to make an informed choice" (Brannigan et al., 2011). Subsequently, adopting to the vehicle labelling directive, the German Energy Agency encouraged auto buyers to check $CO_2$ emission data on new vehicles through the "Passenger Vehicle Energy Consumption Labelling" (Pkw-EnVKV) which was officially passed in 2004 (Kotak & Kotak, 2016).

# References

Ahuti, S. (2015). Industrial growth and environmental degradation. *International Education & Research Journal, 1*(5), 5–7.

Apte, S., & Sheth, J. (2017). Developing the sustainable edge. *Leader to Leader, 2017*(85), 48–53. https://doi.org/10.1002/ltl.20306

Arowoshegbe, A. O., & Emmanuel, U. (2016). Sustainability and triple bottom line: An overview of two interrelated concepts. *Igbinedion University Journal of Accounting, 2*, 88–126.

Artsiomchyk, Y., & Zhivitskaya, H. (2015). Designing sustainable supply chain under innovation influence. *IFAC-PapersOnLine, 48*(3), 1695–1699. https://doi.org/10.1016/j.ifacol.2015.06.330

Bansal, P. (2005). Evolving sustainably: A longitudinal study of corporate sustainable development. *Strategic Management Journal, 26*(3), 197–218. https://doi.org/10.1002/smj.441

Beske, P., Land, A., & Seuring, S. (2014). Sustainable supply chain management practices and dynamic capabilities in the food industry: A critical analysis of the literature. *International Journal of Production Economics, 152*, 131–143. https://doi.org/10.1016/j.ijpe.2013.12.026

Boström, M., Jönsson, A., Lockie, S., Mol, A., & Oosterveer, P. (2015). Sustainable and responsible supply chain governance: Challenges and opportunities. *Journal of Cleaner Production., 107*, 1–7. https://doi.org/10.1016/j.jclepro.2014.11.050

Brandenburg, M., & Rebs, T. (2015). Sustainable supply chain management: A modeling perspective. *Annals of Operations Research, 229*(1), 213–252. https://doi.org/10.1007/s10479-015-1853-1

Brannigan, C., Kay, D., Gibson, G., & Skinner, I. (2011). *Report on the implementation of directive 1999/94/EC relating to the availability of consumer information on fuel economy and CO2 emissions in respect of the marketing of new passenger cars.* https://ec.europa.eu/clima/sites/clima/files/transport/vehicles/labelling/docs/final_report_2012_en.pdf

Brown, M. (2017). *What are the benefits of sustainable manufacturing?* https://www.cadcrowd.com/blog/what-are-the-benefits-of-sustainable-manufacturing/

Carter, C. R., & Easton, P. L. (2011). Sustainable supply chain management: Evolution and future directions. *International Journal of Physical Distribution and Logistics Management, 41*(1), 46–62.

Carter, C. R., & Rogers, D. S. (2008). A framework of sustainable supply chain management: Moving toward new theory. *International Journal of Physical Distribution and Logistics Management, 38*(5), 360–387.

Clemens, B., & Douglas, T. (2006). Does coercion drive firms to adopt 'voluntary' green initiatives?: Relationships among coercion, superior firm resources, and voluntary green initiatives. *Journal of Business Research, 59*(4), 483–491.

Deloitte. (2018). *Global automotive consumer study 2018*. https://www2.deloitte.com/nl/nl/pages/consumer-industrial-products/articles/global-automotive-consumer-study-2018.html

Elkington, J. (1998). *Cannibals with forks: The triple bottom line of 21st century business. Conscientious commerce.* New Society Publishers.

Esty, D. C., & Winston, A. S. (2009). *Green to gold: How smart companies use environmental strategy to innovate, create value, and build a competitive advantage* (Rev. and updated ed.). Wiley; John Wiley [distributor].

EUR-Lex. (2020). *Environment and climate change*. https://eur-lex.europa.eu/summary/chapter/environment.html?root_default=SUM_1_CODED%3D20,SUM_2_CODED%3D2001&locale=en

European Commission. (2019). *Ecolabel: Facts and figures*. https://ec.europa.eu/environment/ecolabel/facts-and-figures.html

European Commission. (2020). *EU Emissions Trading System (EU ETS)*. https://ec.europa.eu/clima/policies/ets_en

Figge, F., Hahn, T., Schaltegger, S., & Wagner, M. (2002). The sustainability balanced scorecard: Linking sustainability management to business strategy. *Business Strategy and the Environment, 11*(5), 269–284.

Freeman, R. E., & Evan, W. M. (1990). Corporate governance: A stakeholder interpretation. *Journal of Behavioral Economics, 19*(4), 337–359. https://doi.org/10.1016/0090-5720(90)90022-Y

Friedrich Ebert Stiftung. (2015). *The future of the German automotive industry: Structural change in the automotive industry: challenges and perspectives. WISO Diskurs: Vol. 2015,20.* Friedrich Ebert Stiftung.

Frondel, M., Schmidt, C. M., & Vance, C. (2011). A regression on climate policy: The European Commission's legislation to reduce $CO_2$ emissions from automobiles. *Transportation Research Part A: Policy and Practice, 45*(10), 1043–1051. https://doi.org/10.1016/j.tra.2009.12.001

Fullerton, K. (2017). *German government offers financial incentives for electric cars*. https://greenbuzzberlin.de/german-government-offers-financial-incentives-electric-cars/

Gimenez, C., Sierra, V., & Rodon, J. (2012). Sustainable operations: Their impact on the triple bottom line. *International Journal of Production Economics, 140*(1), 149–159. https://doi.org/10.1016/j.ijpe.2012.01.035

Goel, P. (2010). Triple bottom line reporting: An analytical approach for corporate sustainability. *Journal of Finance, Accounting, and Management, 1*(1), 27–42.

Green Business Bureau. (2019). *Financial benefits of an eco-friendly business.* https://greenbusinessbureau.com/blog/financial-benefits-of-an-eco-friendly-business/

Green, K. W., Zelbst, P. J., Meacham, J., & Bhadauria, V. S. (2012). Green supply chain management practices: Impact on performance. *Supply Chain Management: An International Journal, 17*(3), 290–305. https://doi.org/10.1108/13598541211227126

Güner, S., & Coskun, E. (2010). The roles of customer choices in green supply chain management: An empirical study in Sakarya region. 8th International Logistics and Supply Chain Congress 2010, Istanbul, Turkey.

Hahn, T., & Figge, F. (2011). Beyond the bounded instrumentality in current corporate sustainability research: Toward an inclusive notion of profitability. *Journal of Business Ethics, 104*(3), 325–345. https://doi.org/10.1007/s10551-011-0911-0

Handfield, R. B., Ragatz, G. L., Petersen, K. J., & Monczka, R. M. (1999). Involving suppliers in new product development. *California Management Review, 42*(1), 59–82.

Handfield, R. B., Sroufe, R., & Walton, S. (2005). Integrating environmental management and supply chain strategies. *Business Strategy and the Environment, 14*(1), 1–19.

Hervani, A. A., Helms, M. M., & Sarkis, J. (2005). Performance measurement for green supply chain management. *Benchmarking: An International Journal, 12*(4), 330–353.

Hsu, C. C., Tan, K.-C., Zailani, S. H. M., & Jayaraman, V. (2013). Supply chain drivers that foster the development of green initiatives in an emerging economy. *International Journal of Operations & Production Management, 33*(6), 656–688.

Hu, A. H., & Hsu, C.-W. (2010). Critical factors for implementing green supply chain management practice. *Management Research Review, 33*(6), 586–608. https://doi.org/10.1108/01409171011050208

Hunke, K., & Prause, G. (2014). Sustainable supply chain management in German automotive industry: Experiences and success factors. *Journal of*

*Security and Sustainability Issues, 3*(3), 15–22. https://doi.org/10.9770/jssi.2014.3.3(2)

IEA. (2020). *Global EV outlook 2020.* https://www.iea.org/reports/global-ev-outlook-2020

ISO. (1996). *ISO 14001: Environmental management systems – Specification with guidance for use.* https://www.iso.org/standard/23142.html

Jantan, M., Ndubisi, N. O., & Hing, L. C. (2006). Supplier selection strategy and manufacturing flexibility: Impact of quality and technology roadmaps. *Asian Academy of Management, 11*(1), 19–47.

Jennings, P. D., & Zandbergen, P. (1995). Ecologically sustainable organizations: An institutional approach. *Academy of Management Review, 20*, 1015–1052.

Jürgens, U., & Meißner, H.-R. (2005). *Arbeiten am Auto der Zukunft: Produktinnovationen und Perspektiven der Beschäftigten.* Ed. Sigma.

Kaplan, R. S., & Norton, D. P. (1996). Linking the balanced scorecard to strategy. *California Management Review, 39*(1), 53–79.

Kemp, R., & Arundel, A. (1998). *Survey indicators for environmental innovation.* https://backend.orbit.dtu.dk/ws/portalfiles/portal/115329898/2007_115_report.pdf

Koh, S. C. L., Gunasekaran, A., & Tseng, C. S. (2012). Cross-tier ripple and indirect effects of directives WEEE and RoHS on greening a supply chain. *International Journal of Production Economics, 140*(1), 305–317. https://doi.org/10.1016/j.ijpe.2011.05.008

Kotak, B., & Kotak, Y. (2016). Review of European regulations and Germany's action to reduce automotive sector emissions. *European Transport, 7*(61), 1–19.

Kumar, A., & Joo, H. (2019). *World biodiesel policies and production* (1st ed.). CRC Press/Taylor & Francis Group.

Leung, T. (2020). *Key advantages of sustainable manufacturing.* http://www.winman.com/blog/key-advantages-of-sustainable-manufacturing

McGill. (2020). *What is sustainability?* https://www.mcgill.ca/sustainability/files/sustainability/what-is-sustainability.pdf

Nielsen. (2018). *Global consumers seek companies that care about environmental issues.* https://www.nielsen.com/eu/en/insights/article/2018/global-consumers-seek-companies-that-care-about-environmental-issues/

Ortas, E., Moneva, J. M., & Álvarez, I. (2014). Sustainable supply chain and company performance. *Supply Chain Management: An International Journal, 19*(3), 332–350. https://doi.org/10.1108/SCM-12-2013-0444

Pagell, M., & Gobeli, D. (2009). How plant managers' experiences and attitudes toward sustainability relate to operational performance. *Production and Operations Management, 18*(3), 278–299. https://doi.org/10.1111/j.1937-5956.2009.01050.x

Pereira de Carvalho, A., Barbieri, A., & Carlos, J. (2012). Innovation and sustainability in the supply chain of a cosmetics company: A case study. *Journal of Technology Management & Innovation, 7*(2), 144–156. https://doi.org/10.4067/S0718-27242012000200012

Porter, M. E. (2008). The five competitive forces that shape strategy. *Harvard Business Review, 86*, 78–93.

PricewaterhouseCoopers. (2020). *The world in 2050: Will the shift in global economic power continue?* https://www.pwc.com/gx/en/issues/the-economy/assets/world-in-2050-february-2015.pdf

Qudrat-Ullah, H. (2018). *Innovative solutions for sustainable supply chains. Understanding complex systems.* Springer.

Rao, P., & Holt, D. (2005). Do green supply chains lead to competitiveness and economic performance? *International Journal of Operations & Production Management, 25*(9), 898–916. https://doi.org/10.1108/01443570510613956

Rebs, T., Brandenburg, M., Seuring, S., & Stohler, M. (2018). Stakeholder influences and risks in sustainable supply chain management: A comparison of qualitative and quantitative studies. *Springer, 11*(2), 197–237.

Richter, W. (2020). *Tesla's global deliveries compared to the top 10: Volkswagen, Toyota, GM, Ford, Honda, FCA, Mercedes… Here's the chart.* https://wolfstreet.com/2020/01/24/tesla-global-deliveries-compared-to-top-10-volkswagen-toyota-gm-ford-honda-fca-mercedes-chart/

Rivera, J. (2004). Institutional pressures and voluntary environmental behavior in developing countries: Evidence from the Costa Rican hotel industry. *Society & Natural Resources, 17*(9), 779–797. https://doi.org/10.1080/08941920490493783

Rodrigues Vaz, C., Shoeninger Rauen, T., & Rojas Lezana, Á. (2017). Sustainability and innovation in the automotive sector: A structured content analysis. *Sustainability, 9*(6), 880. https://doi.org/10.3390/su9060880

Rogers, P., Jalal, K., & Boyd, J. (2007). *An introduction to sustainable development.* Routledge.

Russo, A., Morrone, D., & Calace, D. (2015). The Green side of the automotive industry: A consumer-based analysis. *Journal of Marketing Development and Competitiveness, 9*(2), 59–71.

Saeed, M., & Kersten, W. (2019). Drivers of sustainable supply chain management: Identification and classification. *Sustainability, 11*(4), 1137. https://doi.org/10.3390/su11041137

Sarkis, J., Gonzalez-Torre, P., & Adenso-Diaz, B. (2010). Stakeholder pressure and the adoption of environmental practices: The mediating effect of training. *Journal of Operations Management, 28*(2), 163–176.

Schaltegger, S., & Synnestvedt, T. (2002). The link between 'green' and economic success: Environmental management as the crucial trigger between environmental and economic performance. *Journal of Environmental Management, 65*(4), 339–346. https://doi.org/10.1006/jema.2002.0555

Schwarzkopf, J., & Dorwald, T. (2019). *Self-assessment questionnaire on CSR/sustainability for automotive sector suppliers.* https://drivesustainability.org/wp-content/uploads/2019/06/Supplier-handbook_final-version.pdf

Scott, C., Lundgren, H., & Thompson, P. (2018). *Guide to supply chain management. Management for professionals.* Springer.

Seipp, V., Michel, A., & Siegfried, P. (2020). Review of international supply chain risk within banking regulations in Asia, US and EU including proposals to improve cost efficiency by meeting regulatory compliance. *Journal Financial Risk Management, 9,* 229–251. https://doi.org/10.4236/jfrm.2020.93013

Seuring, S., & Müller, M. (2008). From a literature review to a conceptual framework for sustainable supply chain management. *Journal of Cleaner Production, 16*(15), 1699–1710. https://doi.org/10.1016/j.jclepro.2008.04.020

Siegfried, P. (2014). *Knowledge transfer in service research - Service engineering in startup companies.* EUL-Verlag. ISBN: 978-3-8441-0335-9, https://www.eul-verlag.de/shop/eul/apply/viewdetail/id/2420/

Siegfried, P. (2015). *Die Unternehmenserfolgsfaktoren und deren kausale Zusammenhänge, Zeitschrift Ideen- und Innovationsmanagement* (pp. 131–137). Deutsches Institut für Betriebswirtschaft GmbH/Erich Schmidt Verlag, ISSN 2198-3143. https://doi.org/10.37307/j.2198-3151.2015.04.04

Siegfried, P. (2017). *Corporate strategic management in practice,* ISBN: 978-3-86924-985-8. AVM Akademische Verlagsgemeinschaft.

Siegfried, P., & Strak, D. (2021). *Grüne Logistik: Eine Untersuchung ausgewählter alternativer Antriebstechnologien im Güterverkehr, Zeitschrift für Verkehrswissenschaft (ZfV).* D-Journal. ISSN: 0044-3670.

Siegfried, P., & Zhang, J. (2021). Developing a sustainable concept for the urban last mile delivery. *Open Journal of Business and Management, 9*(1), 268–287. https://doi.org/10.4236/ojbm.2021.91015

Slaper, T. F., & Hall, T. J. (2011). *Triple bottom line: What is it and how does it work?* https://www.ibrc.indiana.edu/ibr/2011/spring/pdfs/article2.pdf

Sroufe, R. (2003). Effects of environmental management systems on environmental management practices and operations. *Production and Operations Management, 12*(3), 416–431. https://doi.org/10.1111/j.1937-5956.2003.tb00212.x

Telang, S. (2013). *Importance of greening automotive supply chain - Green clean guide.* https://greencleanguide.com/importance-of-greening-automotive-supply-chain/

Thompson, S. (2020). *How do consumers affect supply chain management?* https://smallbusiness.chron.com/consumers-affect-supply-chain-management-81664.html

UNFCCC. (1992). *United Nations Framework Convention on Climate Change.* https://unfccc.int/resource/docs/convkp/conveng.pdf

UNFCCC. (2008). *Kyoto protocol: Reference manual on accounting of emissions and assigned amount.* https://unfccc.int/resource/docs/publications/08_unfccc_kp_ref_manual.pdf

Vachon, S., & Klassen, R. D. (2008). Environmental management and manufacturing performance: The role of collaboration in the supply chain. *International Journal of Production Economics, 111*(2), 299–315. https://doi.org/10.1016/j.ijpe.2006.11.030

Vermeulen, W. J. V., Uitenboogaart, Y., Pesqueira, L. D. L., & Metselaar, J. (2011). *Other roles for governments needed in sustainable supply chain management?* https://www.pbl.nl/en/publications/other-roles-for-governments-needed-in-sustainable-supply-chain-management

Virta. (2020). *The global electric vehicle market in 2020: Statistics & forecasts.* https://www.virta.global/global-electric-vehicle-market

Walley, N., & Whitehead, B. W. (1994). It's not easy being green. *Harvard Business Review, 72*(3), 46–52.

Wu, J., Yang, Z., Hu, X., Wang, H., & Huang, J. (2018). Exploring driving forces of sustainable development of China's new energy vehicle industry: An analysis from the perspective of an innovation ecosystem. *Sustainability, 10*(12), 4827. https://doi.org/10.3390/su10124827

Zhu, Q., Geng, Y., Fujita, T., & Hashimoto, S. (2010). Green supply chain management in leading manufacturers. *Management Research Review, 33*(4), 380–392. https://doi.org/10.1108/01409171011030471

Zhu, Q., & Sarkis, J. (2007). The moderating effects of institutional pressures on emergent green supply chain practices and performance. *International Journal of Production Research, 45*(18–19), 4333–4355.

Zhu, Q., Sarkis, J., & Lai, K.-H. (2013). Institutional-based antecedents and performance outcomes of internal and external green supply chain management practices. *Journal of Purchasing and Supply Management, 19*, 106–117.

# 4

# Green Supply Chain Management

## 4.1 Introduction to Green Supply Chain Management

As mentioned in the previous chapters, environmental problems have become a critical topic which leads to changes in the consumers' behaviour towards environmentally friendly products and the establishment of national and global regulations to eliminate ecological challenges. Therefore, "Going Green" has become a priority for organisations. In the context to environmental relations, "green supply chain management" (GSCM) has emerged as the most important approach in the field of environmental sustainability. The purpose of this study is the observation the company's environmental performance. In comparison to SSCM the social aspect is left out here.

There are many literature sources containing definitions of GSCM which have been developed in the last decades. Basically, GSCM has its roots in green components influencing the supply chain management. According to Zhu and Sarkis (2006), GSCM is "required to develop the material flow impact of organizations and their supply chains on the value-added by balancing and controlling the flow of natural environment, taking into account climate change, pollution material, capital, information, and work". The implementation of green management integrates actors along the supply chain from suppliers over manufacturers to the customers. In addition, regulatory requirements from governments are drivers for GSCM. C.-W. Hsu and Hu (2008) defined that "GSCM is as an approach for improving performance of the processes and products according to the requirements of the environmental regulations". Performing GSCM enhances competitive and economic advantages for organisations. Balon et al. (2016) have asserted that it is a verified way to diminish an organisation's impact on ecology, while achieving better production performances.

In comparison to the traditional supply chain management, in the recent years, green supply chain management has been considered as a significant concept for manufacturers on the sustainable development path (Siegfried et al., 2021). While the traditional supply chain focuses on the economy as its main objective, the GSCM concentrates on ecological systems. In their study with 186 participants, Zhu and Sarkis (2004) agreed that GSCM practices activate "win–win" relationships between environmental and economic performances. There are several benefits organisations may attain by the implementation of GSCM such as a reduction of environmental hazards, materials costs towards suppliers, manufacturing costs, costs of ownership for customers and resources consumption (see Fig. 4.1). These benefits delight stakeholders' demands and brighten companies' reputations (Sanket Tonape, 2013).

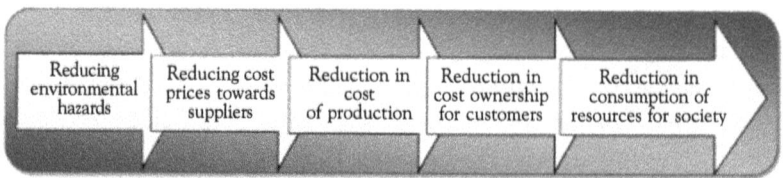

**Fig. 4.1** Benefits of green supply chain management (Sanket Tonape, 2013)

## 4.2 Principles of Implementing Green Supply Chain Management

As stated in the study of Emmett and Sood (2010), the adoption of "Green" in the supply chain management covers upstream and downstream stages within the supply chain that are related to product design, purchasing material, supplier selection, production processes, distribution and disposal recycling. Figure 4.2 illustrates the implementation of "Green" in all stages of the supply chain.

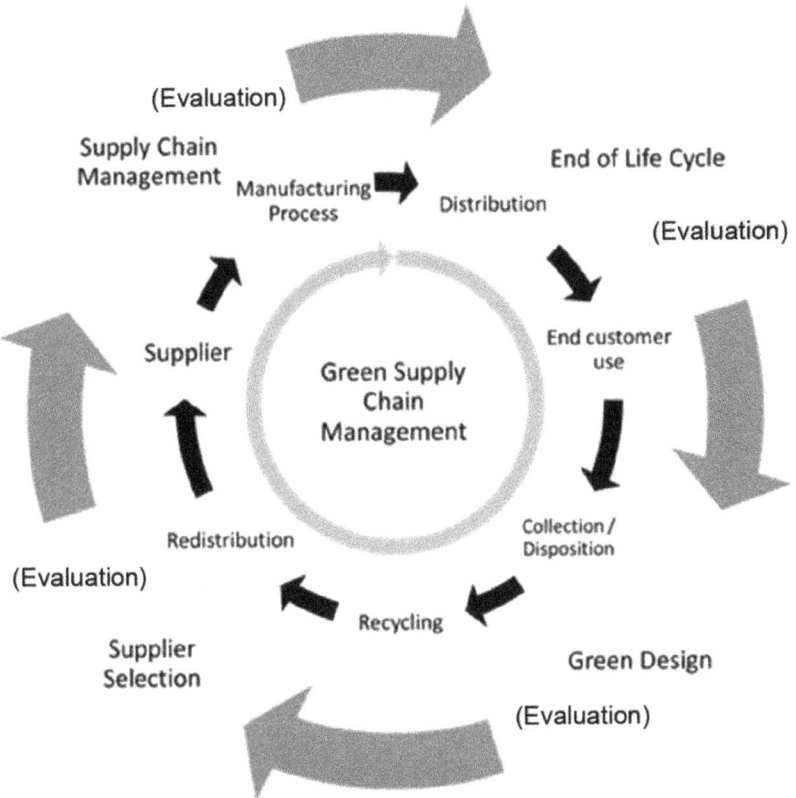

**Fig. 4.2** Green supply chain management process (own illustration based on Diabat & Govindan, 2011)

To effectively implement sustainable development in the supply chain on a daily business, many researches have explored various critical sustainable practices being operated along the supply chain network. These are presented within the next sub-chapters.

## 4.2.1 Green Suppliers Selection

As mentioned before, manufacturers and suppliers have coherent relationships in the automotive industry. The rising percentage of supplier parts in an end-product indicates that a large number of manufacturing steps are outsourced to different tier suppliers in the world (Kallstrom, 2019). Therefore, the supplier selection process is a crucial practice to seek for suitable material and helps the whole supply chain management to operate fluently and effectively. Today, most manufacturers aim to implement sustainability in their own supply chain network and thus in their supplier selection. As reported by Igarashi et al. (2013), supplier selection is seen as one of the most important decision for business operations. The right suppliers will provide the right products with competitive prices (Igarashi et al., 2013).

Different from the traditional supplier selection process which focuses on economic efficiency and ignores ecological impacts, sustainable supplier search should take environmental protection issues into account too (Kumar et al., 2014). Going green has become a main criterion in the purchasing process which is assessed with the ISO 14001 certification for green suppliers (ISO, 2015). In order to extend the green concept within the supply chain effectively it is required to adopt environmental considerations into supplier selection, assessment and collaboration (Gavronski et al., 2011). Selecting green suppliers in the early stage will create better circumstances to implement green practices in further stages of the supply chain. Eco-friendly suppliers play essential roles in establishing green material purchasing. A cooperation with them will not only bring environmental benefits but could also help to build up companies' images in customers' eyes which will affect sales and revenues (Green Purchasing Guide, 2011).

## 4.2.2 Green Product Design

According to Golicic and Smith (2013), eco-friendly design should feature environmental aspects in each stage of the product development and lessen negative effects on the ecology system during the entire product life cycle. Hence, eco-design aims to include environmental thinking into the product's life cycle with its phases from raw material, processing, transportation, distribution and reusing (Srivastava, 2007). Besides, Chen and Sheu (2009) determined that eco-friendly products result in a decrease of waste recycling costs.

Green product design focuses on the following main characteristics (Punit et al., 2015):

- Products made from recycled or remanufactured materials
- Products can be reused and remanufactured
- Products with environment friendly packaging
- Products made from organic components

Furthermore, using energy and resources efficiency plays an essential role in green product design. In 2003, the European Commission first mentioned the energy-using products directive (EuP) for sustainable design practices. The directive consists of ecological requirements concerning efficient energy-using product development which is mandatory within the EU's member states. The goal of the EuP directive is ensuring energy sources availability in the longterm (Grote et al., 2007).

## 4.2.3 Green Material Purchasing

Purchasing is one of the key strategic processes which consist of a series of cooperating activities between many different tier suppliers. Purchasing includes functions like choosing right raw materials, components, parts and supplies which will contribute significantly to manufacturing end-products (Olaore, 2013).

To follow the green concept for the entire business, it is important for manufacturers to practise green purchasing already in the early steps.

According to Min and Galle (1997), green purchasing is understood as an eco-practice that will eliminate the sources of waste and escalate the renewal of materials. The aim of green purchasing is minimisation of negative ecological impacts of manufacturing and transportation processes in the supply chain by using durable, recyclable and reusable materials. Incorporating the green concept in purchasing allows firms to demand from their suppliers to consider green practices (Shao & Ünal, 2019). While price, delivery and quality were the main criteria for purchased materials in the past, today the environmental aspects of the purchased material characteristics are necessary to be considered too. Practising environmental strategies in purchasing will lead to benefits such as cost savings, better public images and competitive advantage in markets for companies (Wisner et al., 2012).

### 4.2.4 Green Manufacturing

In order to operate green strategies in the supply chain, it is important for companies to focus on practising green manufacturing. Many researches mentioned the importance of green manufacturing. The concept was first presented by Crainic et al. (1993) as a well-researched area for sustainable development in SCM. It is defined as a way to optimise production processes towards environmental consciousness which minimises negative environmental impacts like waste and pollution during the product life cycle including manufacturing, usage and disposal (Li et al., 2010). It requires manufacturers to develop innovative green techniques and to use efficient material inputs.

Green supplier selection and green material purchasing are fundamental to approach green manufacturing. Cooperating with green suppliers and exploiting eco-friendly components in early processes will put less pressure on further production steps. Besides, green product design is also considered as an important attribute of green manufacturing (Seuring & Müller, 2008). To succeed in green design, it is inevitable for manufacturers to cooperate with customer's needs. Therefore, the success of green manufacturing relies on the collaboration with suppliers and customers. It involves various green practices from extracting green materials to using green innovative techniques and designing green products which will decrease the use of hazardous substances in the manufacturing process.

## 4.2.5 Green Distribution

Distribution involves transportation activities from the suppliers over the manufacturers up to the end-customers. It generally encloses the whole distribution process, including order processing, storage and warehousing, packaging and labelling, delivery to the customers and taking back packaging (Seuring & Müller, 2008). Distribution of products causes substantial problems for ecological system like carbon dioxide emissions, the greenhouse effect, air pollution or oil spills. According to Sarkis (2006), green distribution integrates environmental concepts into the traditional distribution to eliminate the logistical impact occurring during material flows and product transportation (Michel & Siegfried, 2021).

Basically, sustainable distribution considers creating low negative environment impacts by reducing greenhouse gases and fossil fuel usage (Svensson, 2007). There are many effective ways to approach sustainable transportation such as using green vehicles or public transportation services. Moreover, using hydrogen fuels instead of gasoline or diesel on shipping logistics fleets will help reducing $CO_2$ emissions. Parallel to the development of innovative technologies, electrification can be included into transportation itself, for example, by using energy-efficient e-trucks. In addition, renewable energy sources like solar technology and waste recycling programmes are possible environmental solutions in warehouses (Sustainable Business Toolkit, 2015). Green packaging is also a part of distribution. Size, shape and horizontal or vertical packaging are characteristics impacting on the whole distribution process. Lakshmimeera and Palanisamy (2013) verify that appropriate packaging can lead to reduced material usage and required handling. Optimal packaging facilitates the intensity of space usage in the warehouse. Furthermore, route planning is an essential factor to practise green distribution. McKinnon (2005) declares that direct tour planning, full truck loading and less empty shipments are performance outcomes to reach green distribution.

### 4.2.6 Reverse Logistics

Reserve logistics is understood to be the opposite to traditional logistics. Products are shipped back from customers to manufacturers. According to the definition of Tandem Logistics (2020) reverse logistics is "a specialized segment of logistics focusing on the movement and management of products and resources after the sale and after delivery to the customer". Generally, reverse logistics includes all activities of traditional logistics but implementing it in the reverse way (Hawks, 2006). The purposes of reverse logistics are reducing negative environmental impacts by recycling, reusing, remanufacturing, repairing, refurbishing and disposal of the shipped products from the point of consumption back to the point of manufacturing (Autry, 2005).

The amount of waste in the industries has been increasing rapidly in the recent years which leads to a lot of pressure on the ecology. Therefore, incorporating reverse logistics into the businesses will enhance competitive advantages by improving profitability for manufacturers through cost savings and higher customer satisfaction (Kannan et al., 2009). Reverse logistics affects manufacturers' reputations and brand recognition. Not fulfilling customers' requirements or defected parts are reasons for product returns. In the SCM, reserve logistics refers not only to products returning to manufacturers but also to parts returning from manufacturers to suppliers (Figenbaum & Thomas, 1986).

## 4.3 Benefits of Applying Green Supply Chain Management in the Automotive Industry

Although challenges and barriers are still existing in the implementation of green initiatives into the supply chain management, automotive OEMs have obtained many significant achievements in environmental and financial benefits.

## 4.3.1 Benefits from Environmental Performance

"Going Green" has emerged prominently as a main concept which covers all phases of the product's life cycle from design, production and distribution phases to the use of products by the end consumers and its disposal at the end of the product's life (Borade & Bansod, 2007). Many studies have proven that practising sustainability in the supply chain will pay off for companies. These results are shown by bringing positive effects on the ecological system through $CO_2$ reduction targets of companies (Carbon Disclosure Project, 2011). According to Dekker et al. (2012), green supply chain management is using right materials, right fuels, right technologies and right transportation to prevent environmental burdens. The rise of energy costs, regulation concerning environmental protection and the green demands of customers are factors that require automakers and their suppliers to reduce the carbon footprint of their entire operations, including supply networks.

The German automotive industry strived to apply eco-efficient innovative technologies and concepts in their supply chain to improve the environmental performance of their supply networks.

### 4.3.1.1 Green Material Sourcing

Many of the huge numbers of different materials, needed to produce a car, are traditionally made from steel and aluminium which emits contaminant waste during its production process. Each year the automotive industry uses around 150 million tons of these materials. So, following green strategies, manufacturers are successfully on the path of researching alternative renewable material sources like plant, cotton or coconut fibres which are used as replacements for conventional materials. This achievement enhances the comfort, safety and durability of vehicles while not creating harmful impacts for the environment (Carbon Disclosure Project, 2011). For example, BMW uses carbon fibre that is produced using only hydroelectric power (Cooper-Searle, 2017). Moreover, using lightweight materials instead of traditional steel car frames will result in 50% less weight. The less heavy a car is, the more energy can be saved. Lightweight materials are a great invention for increasing vehicle efficiency and reducing the $CO_2$ emission (Office of Energy Efficiency & Renewable Energy, 2020). The German automotive industry has set up a target for increasing the usage of lightweight materials from 30 to 70 percent by 2030. The Open Hybrid LabFactory was established to research and develop new renewable materials and production techniques to support producing vehicles more environmentally friendly (Bundesministerium für Bildung und Forschung, 2020).

### 4.3.1.2 Green Supplier Selection

Manufacturers and suppliers have tight relationships in the automotive industry. Suppliers play critical roles for OEMs on the way to seek for sustainable development. To implement a green supply chain, many front runner automotive OEMs have applied environmental criteria in their suppliers' network. They require certified environmental management systems such as ISO 14001 and EMAS for their direct suppliers. Reducing waste and increasing recycling become collaborating targets for OEMs and their suppliers to obtain environmental improvement goals (Gaudillat et al., 2017). For example, VW has supported their suppliers to practise EMS for reducing $CO_2$ emissions. Moreover, BMW has also ensured its sustainability standards by setting up supplier self-assessment questionnaires. There are 1900 suppliers' locations to be assessed based on sustainable requirement standards which have been developed by the European Automotive Group on Supply Chain Sustainability (Gaudillat et al., 2017). Suppliers who joined this group reached the target of a reduction of 35 million tons of $CO_2$ emissions. For suppliers who did not fulfil these requirements, OEMs can provide supportive measures and training programmes to encourage environmental performance (Oumer et al., 2015).

### 4.3.1.3 Green Manufacturing

To pursuit green concepts and produce green vehicles, automotive OEMs need to involve environmentally friendly strategies in their automobile assembling process. The main green manufacturing and logistical issues are related to carbon dioxide ($CO_2$) emissions, waste management and water and energy utilisation (Oumer et al., 2015).

As mentioned in Sect. 3.4.2, the environmental management system (EMS) is an aiding tool for companies to approach environment protection measures and to estimate impacts of companies' activities on the environment (Sroufe, 2003). Many global and European OEMs have included EMS in their manufacturing plants. Advanced systems can be implemented according to ISO 14001 or EMAS (EU Eco-management and audit scheme). These systems determine mechanisms for companies to maintain and improve their environmental performance in the highest standards in the long term. German automakers have directed these standards as solutions for going green on their production sites. For example, BMW has implemented systems based on ISO 14001 and EMAS for all production locations throughout the world to reduce their resources and energy consumption, $CO_2$ emissions and process of wastewater significantly (Gaudillat et al., 2017). Moreover, many technological solutions have been given in the manufacturing process for the reduction of greenhouse gas emissions, energy and water conservation, waste management and recycling. For example, paint shops have been converted with waterborne types and water-based solvents in order to reduce environmental impacts during the production processes (Nunes & Bennett, 2010).

### 4.3.1.4 Green Distribution

Distribution also plays an important role in the supply chain to make companies greener. As defined in the former chapter, distribution processes can cause negative effects on the ecological system like carbon dioxide emissions, air pollution or oil spills. Planning green distribution integrates distributing activities in effective ways such as using green vehicles and public transportation services or optimising transportation routes in order to reduce impacts on the ecology (Svensson, 2007).

German automakers have preferred utilising different types of transportation means like giga-liners (long trucks), extra-long trains and larger vessels at sea to decrease the emissions per kilo transported (Hunke & Prause, 2014). BMW has succeeded in practising green distribution in its supply chain. To keep $CO_2$ emissions low, half of the new produced cars of BMW have been transported by railway networks (BMW Group, 2018).

### 4.3.1.5 Green Recycling

The automotive industry is a resource-intensive industry. Therefore, it is necessary to eliminate waste during supply chain networks for preventing negative impacts on the environment. In the European Waste Framework Directive, the waste management plans have been described for OEMs to help them setting up waste minimisation targets (see Fig. 4.3).

This figure represents the various hierarchy options for treating waste in vehicle manufacturing processes. Many OEMs have achieved that their waste management reaches "zero waste to landfill" and "zero waste to incineration" objectives by moving up the waste management hierarchy. The priority in waste management is "Reduce" which targets on reducing waste in the whole production by applying optimal plans and methods. "Re-use" refers to the reuse of materials to extend the product's life before it becomes waste. "Recycling" is the recovery of wasted materials into new products. "Recovery" refers to using new green technologies to generate energy in forms of electricity or heat from direct combustion of waste in cases where the generated waste cannot be reused or recycled.

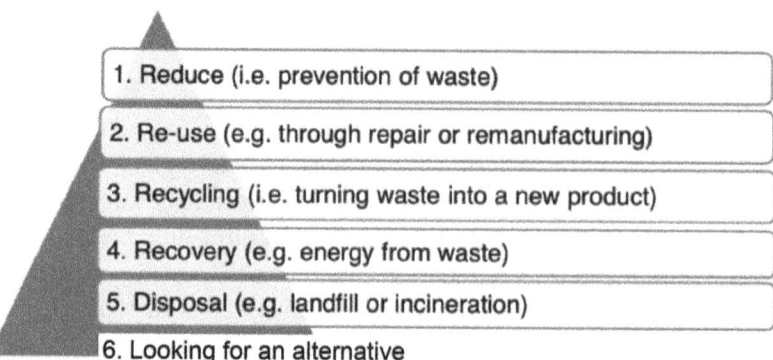

**Fig. 4.3** The hierarchy of options for the treatment of waste during the manufacture of vehicles (own illustration based on Gaudillat et al., 2017)

"Disposal" addresses incentives for applying waste hierarchies such as landfill or incineration charges (Gaudillat et al., 2017).

In 2012, Volkswagen reduced the width of steel used to make parts for the car body as a method to optimise material utilisation. It leads to the advantages that the new Golf produces 15% less waste during its production compared to its predecessor. Besides, by using waste recycling techniques, disposing debris from carbon fibre reinforced plastic (CFRP) components can be returned to the production process of BMW (Gaudillat et al., 2017). Reverse logistics has become a commonly used concept to reach a green supply chain for automotive OEMs. They have realised that the effective reverse logistics could improve companies' outcomes. Remanufacturing, recycling and reusing the disposal of products can reduce negative environmental impacts (Autry, 2005).

### 4.3.1.6 Green Product Design and Green Mobility Development

The trend of a higher environmental perception leads customers to choose preferably eco-friendly products which forces OEMs to bring more green products to the market (Abdul-Muhmin, 2007). Therefore, automotive manufacturers have developed electric vehicles as a solution of battling climate changes due to their low levels of emissions. To encourage the green mobility production, the government has imposed many legislations and programmes. For example, the National Platform for Electric Mobility (NPE) describes the policies and measures to increase the number of electric cars (Verband der Automobilindustrie, 2010). The aim of the National Electromobility Development Plan is to have one million electric vehicles on the roads by 2020 (Germany Trade & Invest, 2015).

Due to the encouraging incentives of environmental legislations from governments and the awareness of consumers on global warming, many automotive pioneers have invested in their technologies for developing electric vehicles. In the last decades, electric vehicles have rapidly evolved in the automotive sector (Singh, 2020). The electric vehicle market has reached 3.2 million units in 2019 and is estimated to grow to more than 26.9 million units by 2030. The number of sales in 2019 was valued at

$162.3 billion in revenues for the global automotive sector and is estimated to reach $802.1 billion by 2027. The registering compound annual growth rate (CAGR) is at 22.6% (Singh, 2020).

Unlike using internal combustion engines, electric vehicles are operated using an electric motor which requires batteries to deliver energy to the engine. Different to the conventional gasoline or diesel cars, EVs are designed to reduce $CO_2$ emissions which lead to environmental pollution. To follow the new trend of replacing combustion engine cars by electric engine cars, automotive players like Tesla (US), BYD (China), BMW (Germany), Volkswagen (Germany) and Nissan (Japan) have adopted new green strategies and technologies and expanded collaboration with partners in order to become global dominants and to gain high growths from electric vehicles (McKinsey, 2020).

Today there are three types of electric vehicles: battery electric vehicles (BEV), hybrid electric vehicles (HEV) and plug-in hybrid electric vehicles (PHEV). EVs use a battery pack with thousands of Li-ion cells working together. When the car is running, the chemical reaction inside the battery creates electricity that powers the motors and so the car's wheels (EDF Energy, 2020). Although EVs meet with environmental protection regulations by their non-emitted $CO_2$ engines, there are conflicts about the usage of the materials, like lithium or cobalt, used in the batteries which may have other negative impacts on the environment and society.

For over 10 years, the demand on lithium-ion battery has constantly increased due to its usage in electric mobiles. Evidently, EVs basically do not emit $CO_2$ during their operation, but the manufacturing process of EVs and lithium batteries itself may cause greenhouse gas emissions (Jenu et al., 2020). Lithium is a critical metal. To create a battery, mining and processing of metals like lithium and copper is required, which can release significant toxic emissions. The production process of batteries shows a complicated value chain. It requires parts from more than 20 different materials from various mining locations around the world, which include many refining and extracting stages. Therefore, the production process of batteries is energy-intensive which results in different climate impacts (Jenu et al., 2020).

Furthermore, the battery industries are facing problems of child and forced labours from local mining procurement. For instance, in the

Democratic Republic of Congo, many cobalt operations have been involved in scandals of forcing children to mine the mineral which is used in batteries as well (Götze, 2019). Therefore, to improve the environmentally friendly image and social responsibilities, it is important for automakers to verify the sources and mining methods from their procurement process in the supply chain. With the new plans of launching more electric cars, Volkswagen has verified new rules to all its suppliers and subcontractors "to ensure there's no child labor in the supply chain" (Thompson, 2018). Daimler has required new processes for suppliers "to disclose their supply chain right through to the mine" (Petroff, 2018). BMW has considered buying "the prized mineral directly from miners to avoid operations that exploit children" (Petroff, 2018).

### 4.3.2 Benefits from the Economic Performance

As environmental practices have attained significant achievements in the supply chain of the automotive industry, this leads to improvements of economic benefits for OEMs.

Many studies have clarified that the $CO_2$ reduction is the fundamental target for sustainability in the supply chain. This target brings better financial performances to companies by getting higher return on investments and this is a powerful tool for costs cutting (Hunke & Prause, 2014). The Carbon Disclosure Project (CDP) is an organisation based in the UK to support cities and companies in disclosing environmental impacts. It aims to increase awareness and actions towards ecological protection and sustainable economy development. According to a survey of CDP, it is explored that more than half of the participating companies and suppliers experienced cost reductions as they practised sustainable supply chain activities (Hunke & Prause, 2014).

Applying sustainable concepts is a conventional way for business development. Inventing innovation technologies based on environmental standards helps to generate economic results. For example, using environmental management systems (EMS) results in financial advantages like cost savings from consuming fewer resources, producing less waste, operational efficiencies and reduced liabilities (Commission for Environmental Cooperation, 2005).

The increase of costs due to the scarcity and volatility of raw materials leads to higher investments in energy saving equipment (Gaudillat et al., 2017). Since that, automotive companies can lower their energy demand significantly. Besides, prevention of waste helps companies to achieve cost savings by using less raw materials and by avoiding more disposal processes. The implementation of a green supply chain based on using green materials, green manufacturing and recycling processes will achieve economies of scale for automakers through reduction in material and energy consumption (Gaudillat et al., 2017). The green production leads to savings in materials because of reuse and recycling. The lower costs occurring in the supply chain will bring automakers more competitive advantages and better opportunities to win over competitors through their efficiency (Gaudillat et al., 2017).

Furthermore, applying sustainable concepts into the organisation is a strategy for automakers to attract more customers who care about eco-friendly products. There are studies on consumer behaviour which are based on a sample of 7700 participants that have proven that sustainable companies can generate more turnovers and revenues from customer sales transactions due to their green images and reputations (Hunke & Prause, 2014). Moreover, governments give many subsidies and economic incentives for actors who choose environmentally friendly solution in their supply chain (Gaudillat et al., 2017). These economic benefits create more motivations for OEMs to practise sustainable concept in their supply chain.

## 4.4 Barriers and Challenges of Green Supply Chain Management in the Automotive Industry

Forcing factors such as higher energy costs, regulation concerns and changing demands of customers require automakers and their suppliers to reduce the carbon footprint of their entire operations, including the supply networks. Regarding the "green" challenges, the focus on the environment might reshape this supply chain scenario even more radically.

The increasing governmental regulations with respect to environmental standards will raise costs but also increase complexity for the automotive industry. Carbon dioxide regulations have mostly been imposed to reduce emissions. Many countries like the USA, Japan, China and European states have enacted to these laws through investing in green supply chain management. The problems for these industries are the initial large investments to become greener (Pereseina et al., 2014). For example, in Europe, it is targeted to reduce $CO_2$ with the help of advanced technologies in new produced vehicles. Electrification could be the key to solve environmental problems. This will push OEMs to invest more in e-mobility vehicles which use electrical/hybrid power-trains and batteries, as well as in lightweight technologies.

However, investments in green technologies result in higher costs (McKinsey, 2020). These costs may inhibit environmental goals of automotive companies. Green supply chain management is a new concept which just started to be popular in the last decades so that it still lacks acceptance of companies due to the increased costs of investments. The companies are uncertain if they can get any economic benefits returning from the advanced green investments. Investing without returns is not a normal thing to do for a business while trying to survive in a long term (Pereseina et al., 2014). Although many automotive companies want to practise GSCM, they are facing the issue of balance between being environmentally friendly and protecting the nature, on the one hand, and meeting the needs of profit-hungry shareholders, on the other hand (Gifford, 1997). The study of Walker et al. (2008) shows that cost concerns might be the most serious barrier for considering environmental factors in the purchasing process. It puts automotive companies under the mindset of ecology versus economy trade-off. The difficulties for the companies are how to practise social and environmental performances together with economic ones (Pereseina et al., 2014).

Besides, according to Chan and Kumar (2007), market demand is the most important external barrier on practising GSCM. Customers who desire lower prices of vehicles will become obstacle factors that prevent automotive maker from investing in green technologies. Generally, customers may not prefer purchasing green products willingly or pay more to obtain additional services (Kuo et al., 2010).

As the automotive industry extended globally with production facilities in different countries, it created more complexities for automotive companies to reshape the supply chain networks into green management. Outsourcing in the automotive industry requires a long supply chain network and partners from many countries. It results in varieties of different environmental acts and regulations that can cause the inhomogeneity in practising green strategies between actors in the supply chain (Xia & Tang, 2011). On the other hand, the long supply chain networks cause an intense bullwhip effect, the uncertainty of demands. It wastes a huge amount of financial resources in inventory management for automakers. Furthermore, transportation costs can be extremely high and transportation capacities along with accidents such as natural disasters can severely hurt the stability of the supply chain. Therefore, many opportunities to adjust the supply chain into being green are lost because of the long supply chain (Xia & Tang, 2011).

The authors D. S. Rogers and Tibben-Lembke (1999) asserted that the lack of top management commitments is seen as the crucial barrier to the successful implementation of green supply chain (Balon et al., 2016). Due to the high costs on green supply chain investments and the slow rate of returns, managers of automakers put a higher priority on economic preferences over environmental performance (Balon et al., 2016). The implementation of green supply chain relating to reverse logistics is challenging for OEMs in both ecological and economical aspects. Although there is a lot of evidence that designing reverse logistics is the tool to save operational costs and to improve the profitability and customers' satisfaction, the lack of knowledge about reverse logistics as well as green concepts may lead to unsuccessful green supply chain implementation (Kannan et al., 2009).

The automotive industry consists of multiple supplier tiers. Such that, the lack of eco literacy of suppliers becomes one of the barriers of GSCM. The cooperation in ecological concerns between different suppliers and partners in different levels of supply chain networks is fundamental for OEMs to practise GSCM. Without awareness of GSCM among suppliers, the OEMs may lose their competitive advantages and ruin their green images and reputations (Balon et al., 2016).

To practise GSCM, the OEMs need to update their new green innovative technologies in their supply chain. Hence, the training programme for their workforce is necessary to raise their working performance level which may result in efficient and profitable GSCM implementations. However, the automakers are still facing the resistance to change and to adopt innovation from their employees. Although GSCM is becoming more prevalent in Europe and USA markets, in developing countries there is still a lack of supportive laws and regulations to support OEMs in the implementation of GSCM (Muduli et al., 2013). Support from the authorities is essential to minimise international misunderstandings and the incentives from governments give more budgets to automakers for investing in green technologies (Chandramowli et al., 2011).

## 4.5 Case Studies: German Automotive Companies Using Green Supply Chain Management

This chapter will analyse the orientation and implementation of sustainable and green supply chain management of three big leading automotive manufacturers in Germany, Volkswagen, Daimler and BMW in details. It also benchmarks the sustainable development to OEMs in Asian markets.

### 4.5.1 German Automotive Industry

With more than 130 years of development history, nowadays, automotive has become the largest industry sector in Germany (Germany Trade & Invest, 2018). Compared to rival OEMs in other countries, German auto manufacturers are recognised as the global leaders. 16.4 million vehicles were produced in Germany in 2007. The German automotive industry has recorded total revenues of 430 billion in 2017. That accounts for 20% of the total domestic industry revenue. Due to the significant growth of the automotive industry, Germany has built up more than 40 OEM sites which are seen as the largest concentration of OEMs in

Europe. Thus, the automotive industry contributes to employment opportunities of around 300.000 people and to the prosperity of the country's economy (Germany Trade & Invest, 2018).

Furthermore, German cars are well-known in the world for safety, innovation and reliable design. That makes Germany the leader of the premium car production worldwide. Around 40% of the global premium brand vehicles are manufactured by German OEMs (Germany Trade & Invest, 2018).

In the recent years, there were many new brands of car makers emerging from Asian markets like China and India. That creates a higher competition for the global automotive market. In order to maintain competitiveness, German automotive companies spent almost EUR 39 billion on R&D development in 2016 which accounts for one third of the global R&D expenditures (Bormann et al., 2018). Therefore, Germany is regarded as a competitive and innovative player in the global automotive industry. One of the most important reasons for the success of the German automotive industry is the density of OEMs and suppliers operating actively in this sector. Besides renowned automakers like Daimler, BMW, Porsche, Volkswagen, Opel, etc., Germany is home to 16 of the world's top 100 automotive OEM suppliers like Continental, Bosch, ZF Friedrichshafen, Schaeffler, etc. (Germany Trade & Invest, 2018). As the automotive industry expands globally, the distribution of production and sales of vehicles has risen across various regions of the globe. Today Germany is no longer a classical vehicle-exporting country, but it becomes the heart of a worldwide production network (Bormann et al., 2018). Almost two-thirds of German OEM vehicles were produced outside the domestic market in the last years and China is the most important manufacturing site outside of the country because of its low labour costs, huge workforces and a giant consumption market. The German automotive industry now has over 2000 production plants around the world. The different manufacturing plants have been spread across the world which leads to a higher complexity of supply chain networks in the automotive industry (Bormann et al., 2018).

## 4.5.2 Volkswagen

Volkswagen AG is one of the top three German automakers. The company was established in 1937 by the German government to massproduce a low-priced "people's car" and is headquartered in Wolfsburg. During the long development history, VW has become a giant multinational corporation in charge of designing and manufacturing multiple car and truck brands. Other than automobile production, the company has expanded in other fields such as financial, leasing services and fleet management. Globally, VW is well-known with its sub brands like Audi, SEAT, Porsche, Lamborghini, Bentley, Bugatti, Scania, MAN and Škoda. In 2016 and 2017, VW has been recognised as the largest automaker measured by worldwide sales. Nowadays, the VW group has locations worldwide in 150 countries and operates 94 production facilitates (Volkswagen AG, 2020a).

As the result of the fraud on diesel emissions in 2015, VW has recalled 8.5 million cars in Europe, including 2.4 million in Germany, 1.2 million in the UK and 500,000 in the USA (Hotten, 2015). This scandal has cost the company EUR 27.4 billion in penalties and fines (Schwartz, 2018). After the failure in this diesel scandal, VW has strived to improve its sustainable image and reputation by pushing their investments on green technologies higher compared to their rivals. The company has planned to spend EUR 20 billion on electrification and EUR 14 billion on shared mobility and self-driving technologies by 2025 (McGee, 2018).

In the sustainability report of 2019, VW has prioritised its future goals to make mobility become more sustainable for future generations (Volkswagen AG, 2020b). Within taking responsibility on the environment, VW wants to develop its entire organisation as carbon neutral by 2050. The decarbonisation programme is the key for developing VW's business model. In the Paris Agreement the company has committed to take the pioneer role in sustainability. VW wants to improve its passenger cars' total lifecycle carbon footprint by 30% compared to 2015. Electrification of its vehicle fleet is the main measure leading to successful sustainable development. 70 new electric models are intended to launch in the market in the next 10 years which will rise the proportion of electric cars in the fleet to at least 40% by 2030 (Volkswagen AG, 2020b). In comparison with conventional vehicles, the new electric models show a much lower energy consumption in case of 200.000 km of driving which is depicted in Fig. 4.4.

However, VW understands that electric mobility is only really environmentally friendly when the carbon footprint over the vehicle's entire life cycle is optimised. Therefore, it is essential to cover sustainable targets in the entire supply chain. The future switch to electric mobility shifts emissions away from the usage phase to the production processes and supply chain's activities. The reduction of $CO_2$ emissions in the supply chain hence becomes the strategic focus for the organisation (Volkswagen AG, 2020b).

By complying with $CO_2$ emission standards and regulations of the Paris Agreement, VW has targeted the emissions of all its plants per vehicle to be reduced by 50% by 2025 compared to 2010 (Volkswagen AG, 2020b). The following measures and objectives are targeted and pursued for the $CO_2$ reduction:

### 4.5.2.1 Energy Efficiency in the Production

VW asserts that already at 43 manufacturing sites the electricity to produce the vehicles is coming 100% from renewable energy sources. The

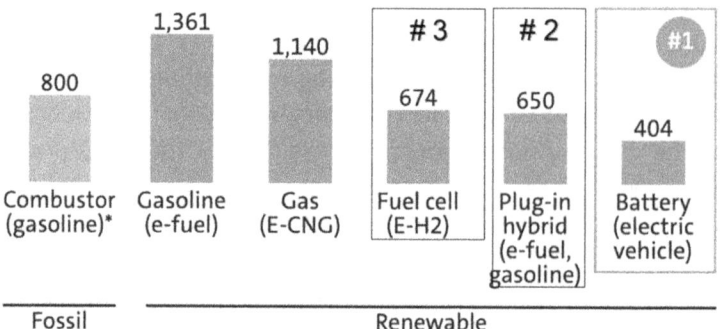

**Fig. 4.4** Primary energy requirement of different vehicle engine types (own illustration based on Volkswagen AG, 2020b)

global proportion of renewable energy in VW reaches to 41% of their electricity consumption (Volkswagen AG, 2020b). The company has invested into the Zwickau factory to become the largest, most efficient and environmentally friendly e-plant in Europe. Purchasing green energy in VW's production sites leads to a $CO_2$ emission reduction of 90,000 tonnes per year. Volkswagen's zero-impact factory initiative contributes to its future goal towards automotive manufacturing with zero impact on the environment. It is part of their environmental mission statement "goTOzero" (Volkswagen AG, 2020b).

### 4.5.2.2 Renewable, Recycled Materials

Replacing materials with renewable raw materials such as natural fibres, cotton, wood results in a better environmental performance by lowering resource consumptions. Aluminium is an important component used in the body construction. The aluminium industry requires high energies in the production process (Volkswagen AG, 2020b). To solve this issue, Audi has implemented the innovative recycling concept "Aluminium Closed Loop" that is aimed to build up a closed-loop recycling system between the company and its suppliers. The rest parts, after cutting the aluminium sheets in Audi's press shops will be returned to suppliers who recycle them into secondary raw materials. It delivers advantages for the environment as the need for energy-intensive production of new aluminium is eliminated (Volkswagen AG, 2020b).

### 4.5.2.3 Recycling Batteries of E-Vehicles

Understanding the negative issues of lithium batteries, VW has implemented measures to recycle the dead batteries. In Salzgitter, the company has set up a battery recycling facility where more than 3000 batteries will be recycled per year. Besides, VW has invented new raw materials (black powder) that can be extracted for the cathodes of new batteries. This leads to $CO_2$ saving potentials of up to 25% (Volkswagen AG, 2020b).

#### 4.5.2.4 Friendly and Safe Waste Management

VW aims to reduce the quantity of waste in the production process and invest in techniques to reuse high-quality recycled materials. Using digital technologies in all German and European factory plants makes it easier to control the waste management process. Based on the purpose of the zero-impact factory initiative, VW puts effort in avoiding plastic usage in its supply chain process through the project "Zero Plastic Waste Factory". For example, Hannover and Pamplona are the sites being successful in avoiding wastes and in cost savings (Volkswagen AG, 2020b).

#### 4.5.2.5 Sustainable Requirements on Suppliers

VW has applied an evaluation system on the sustainable performance of its suppliers. This is a measure to increase the environmental performance for the whole organisation. VW's suppliers with their production facilities must have environmental certificates in accordance with ISO 14001 or EMAS (Volkswagen AG, 2020b). Information about the energy consumption and $CO_2$ emission should be transparent on the request of the VW group. Before making the quotation, suppliers must fulfil sustainable requirements. In the collaboration with battery cell suppliers, it is necessary for suppliers to indicate transparent supply chains from mining the raw materials to manufacturing the finished product. This process will be repeated every 12 months. To extend the responsibility on ecology systems among the suppliers throughout the supply chain, VW develops numerous workshops, hotspot analysis, roadmaps and trainings with suppliers to pursue the objective of emission reductions. In a questionnaire given to 12.646 suppliers in 2019, 5915 suppliers admitted the improvement of sustainable performance through taking these steps (Volkswagen AG, 2020b).

#### 4.5.2.6 Economic Successes from Practising SSCM/GSCM

In accordance with environmental practices in the supply chain, VW has invested around EUR 26 million in factories using renewable energies and EUR 15 million in a combined heat and power plants. There are planned investments of EUR 33 billion in electric mobility development across the group by 2024. Although sustainable development costs for VW are huge investments, they benefit from economies of scale.

VW asserts that investments open chances to achieve profitable growth and strengthen their competitiveness. Besides, practising sustainable performances helps to build up the economic power and profitability of the Volkswagen Group. With the electric vehicles under VW's brand, the company wants to achieve a high market coverage. Green concepts lead to consistent trust and loyalty of customers in VW's brands. In the European markets, Volkswagen passenger cars have achieved stable scores for brand image and brand trust while Porsche is continuously staying on the top position in the image ranking. In 2019, the company has attained a customer satisfaction rate of 83% and is on the way to target a 90% rate in 2025 (Volkswagen AG, 2020b).

### 4.5.3 BMW

Recognised as a top market leader in the premium car segment in the German Automotive industry, BMW has focused its business development on approaching sustainability topics in the passenger car industry as well as on usage of electromobility in the future.

BMW was early established in 1916 in Munich, Bavaria, Germany. Through the long history of its development, BMW still remains a top premium car leader not only in Germany but worldwide. In terms of sales, BMW has recorded 2,520,357 vehicles sold in 2019 (BMW Group, 2020). This showed an increase of 1.2% in comparison to the previous year. Following sustainable strategies, BMW has also strengthened its pioneering market position in the electric vehicle segment. There are around 500.000 e-cars under the BMW brands driving on streets. BMW clarifies that its business bases on sustainable value creation. In the BMW Group Sustainable Value Report 2019 it is stated that sustainable development is the corporate's long-term objective (BMW Group, 2020).

By being aware of the problems of climate change, BMW has set up the sustainability goal on its new vehicle fleet to reduce the $CO_2$ emissions by 50% compared to the year 1995 (BMW Group, 2020). To pursue the goals, BMW has successfully applied these following measures:

### 4.5.3.1 Encouraging Customers with Incentives to Develop Sustainability

By understanding the customers' demands and to meet environmental protection strategies, BMW has given buyers many incentives to encourage them in using more environmental mobility technologies. Examples are premiums and reduced taxes which also play an important role in the buying decisions of customers (BMW Group, 2020).

### 4.5.3.2 Using Technologies as a Solution for Lowering Emissions

Through using efficient technologies, BMW has achieved its sustainable goals of reducing energy consumption and the $CO_2$ emissions. In 2009, the company has launched the 48-volt technology on its BMW 520d and BMW 520d Touring models which can help reducing the fuel consumptions up to 0.3 litres per 100 km. Furthermore, BMW has fitted its diesel vehicle models with new technologies to reduce the emissions of nitrogen oxide ($NO_X$), for example, the $NO_X$ storage catalytic converter or the selective catalytic reduction system (BMW Group, 2020).

### 4.5.3.3 Compliance with Environmental Regulations

The company has supported developing the Worldwide Harmonised Light Vehicles Test Procedure (WLTP) through the VDA (German Association of the Automotive Industry) and ACEA (European Automobile Manufacturers' Association) in order for its vehicle fleets to meet with EU emission targets for 2020 and 2021 (BMW Group, 2020).

### 4.5.3.4 Reducing Emissions in Product Development

By not only paying attention with reducing $CO_2$ emissions in the utilisation of end-products, BMW has taken its environmental performance during the whole product development process. In implementing the life cycle assessment in accordance to ISO 14040/44, BMW has defined its target of emission reduction over the whole vehicle's life cycle in the supply chain from purchasing, production and distribution to vehicle recycling (BMW Group, 2020).

### 4.5.3.5 Raising the Sustainability Awareness in the Supplier's Networks

BMW has required sustainability awareness from its suppliers. Suppliers must make sure to participate in the supply chain programme Carbon Disclosure Project (CDP) for the purpose of $CO_2$ emission reduction. Since 2004, sustainability is an essential purchasing criterion of BMW applied to all suppliers of production materials as well as service providers. Direct suppliers in different tiers need to obligatorily transfer these requirements to their sub-suppliers. In 2019, the participating suppliers in the renewable energy projects due to the energy efficiency increase, have reduced their $CO_2$ emissions by 32 million (BMW Group, 2020).

### 4.5.3.6 Reduction of the Consumption of Resources, Waste and Usage of Renewable Materials

The company aims to decrease $CO_2$ emissions and the energy consumption in the production process by using efficient technologies. As a result, the volume of used resources per vehicle in 2019 has been reduced by 7.8% in comparison to previous years (BMW Group, 2020). This led to economic benefits for the company through savings of EUR 171 million. Moreover, BMW has applied new recycling and reprocessing concepts by sending back material waste during the production to suppliers to reduce the entire waste of the company. The environmental management system of BMW is certified by ISO 14001 and ISO 9001 (BMW Group, 2020).

BMW ensures the usage of eco-friendly raw materials in the early stages of the vehicle production through the programme "Life Cycle Engineering". The company uses renewable materials in its vehicles such as recycled plastics and natural fabrics like flax or kapok. Moreover,

BMW considers to recover vehicles at the end of life as secondary sources of materials (BMW Group, 2020).

Figure 4.5 shows the $CO_2$ emissions per produced vehicle in 2019 and the reduction compared to higher masses in the previous years.

With not only paying attention with using $CO_2$ friendly energy sources, the transport ways used in the distribution are also an essential criterion for BMW to develop sustainability. It shows that more than 50% of BMW's produced vehicles have been distributed to the market by trains. For interplant transportation, electric lorries are used. In addition, BMW has invested in the research and development of biofuel in the sea freight to reduce emissions of sea transports (BMW Group, 2020).

### 4.5.4 Daimler

Like Volkswagen and BMW, Daimler builds sustainability into its business strategy. The Daimler AG defines in its Sustainability Report 2019 that sustainable development is creating a lasting economic value while remaining aware of environmental and social impacts from corporate activities along the entire supply chain (Daimler AG, 2020). The

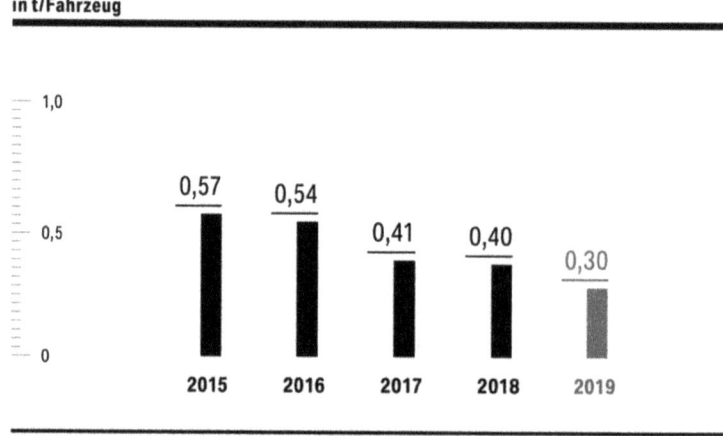

**Fig. 4.5** $CO_2$ emissions per vehicle produced (BMW Group, 2020)

company has committed to sustainable mobility as one of the most important projects to protect the climate, improve the air quality and conserve the natural resources. Therefore, emission-free mobility concepts like electric vehicles are the core element of Daimler's sustainable strategy. The company aims to set up its new fleet to become $CO_2$ neutral by 2039. Climate protective activities will be addressed in all stages of Daimler's supply chain from the material procurement, production, usage phase and recycling and disposal as well as with logistics (Daimler AG, 2020).

### 4.5.4.1 Reducing Material Consumption and Using Closed Material Cycles

The demand on using mobility is increasing globally, hence it leads to the scarcity of resources. The automotive industry requires large amounts of steel, aluminium and polymers in the manufacturing process, this creates negative environmental impacts. To lower the material consumption, Daimler targets to produce its vehicles with new lightweight materials and components. The company is also working with closed material cycles and promotes the use of secondary raw materials in its vehicles (Daimler AG, 2020).

### 4.5.4.2 Working with Suppliers for SSCM

Daimler has approximately around 60.000 direct suppliers from different regions like Europe, North America and Asia. While building up Daimler's supplier network cooperation model, the company has broadened its sustainability guidelines and requirements for its suppliers. It ensures that not only Daimler's direct suppliers but also their upstream value chains commit to sustainability standards. These measures such as screening, risk-based due diligence analyses and training courses are applied enforcing suppliers to compliance with environmental standards like ISO 14001 or EMAS (Daimler AG, 2020).

### 4.5.4.3 Implementing Recycling Along the Supply Chain

All Mercedes-Benz models are 85% recyclable. Daimler has established the Used Part Centre to ensure the vehicles' components and parts can be reused or resold. Besides, Daimler remanufactures car components to prevent waste and unnecessary consumption. The company has optimised production processes to minimise big amounts of waste. Daimler classifies different types of waste and treats them with specific regulations (Daimler AG, 2020).

### 4.5.4.4 Using Renewable Raw Materials

In replacement for materials which have negative impacts on the environment, Daimler employs a range of renewable raw materials such as hemp, kenaf, wool, paper and natural rubber (Daimler AG, 2020).

### 4.5.4.5 Resource-Efficient Technologies

The company has researched new technologies on electric mobility batteries, so that more energy can be stored without increasing the battery volume. The company has invested in battery production networks and sets itself the environmental goals for the next vehicle generation that battery cells will be produced from renewable energy sources (Daimler AG, 2020).

### 4.5.4.6 Production Plants with Climate Protection Goals

Daimler has striven to reduce $CO_2$ emissions in its manufacturing plants. The target is that by 2039, all of Daimler's worldwide plants reach a $CO_2$ neutral production. Starting from 2022, the company plans for all Daimler locations in Germany to produce with electricity from 100 percent renewable sources (Daimler AG, 2020).

### 4.5.4.7 Optimising Transport Logistics

Daimler builds up and connects more transportation hubs together to reduce the distances and set up the goal for using capacities more efficiently. Using innovative transportation concepts with consideration of sustainability criteria, for example, utilising of green transport modes like low emission trucks or train plays an important role (Daimler AG, 2020).

## 4.5.5 Benchmark Against Automotive Manufacturers in Asian Countries

Since the 1990s, the Asian countries have been rapidly developing their economies through industrialisation. The automotive industry has become the strategic factor for Asian countries to be promoted (Abrenica, 1998). The following statistic in Fig. 4.6 describes the sales volume of car leaders in Asian markets in 2019. Japan, China and India are the top three potential growing markets for the automotive industry.

In recent years, many lower-cost brands have risen in emerging markets such as Geely from China and Tata Motors from India (Siegfried, 2021). They are working intensively to compete in global rival markets with other OEMs from developed countries. Over the previous years, China has surpassed the USA by becoming the largest passenger car manufacturer in the world. In 2019, 21.3 million cars have been produced in China. That accounts for almost one third of the world's passenger vehicle production (Wagner, 2020b). According to the European Automobile Manufacturers Association (ACEA), the European automakers are in a particularly difficult position. From 2007 to 2012, new car registrations

**Fig. 4.6** Number of passenger cars sold in the Asia Pacific region 2019, by country or region (own illustration based on Moore, 2020)

decreased across Europe. For example, in Germany they declined by 2% from 3.1 million units, whereas in the French automotive industry they dropped by 8%. In Spain they decreased by 57%, and in Italy by 44% (PricewaterhouseCoopers, 2014).

#### 4.5.5.1 Chinese Automotive Market—Geely's SSCM

In China, the automotive sector plays an important role for the economy development. The industry provides millions of jobs and the Chinese automotive enterprises contribute to around 28% of the worldwide yearly vehicle production in 2018 (Wong, 2020). Parallel to the growth of the vehicle production in China, it fosters negative effects on the ecology system. According to the China Vehicle Environmental Management Annual Report 2017, the emissions from motor vehicles lead to air pollution (BASF, 2020). The energy consumption of automotive industry is extremely high. The report of the Deputy Minister of Ministry of Industry and Information Technology in 2016 states that the vehicle fuel consumption of Chinese automotive industry did not fulfil international standards (BASF, 2020).

Therefore, the tighter pressure from environmental requirements and competition from outside forces the China's automotive industry to follow new transformations. China has set up green developments which will focus on light weighting materials, reduction of fuel consumption and new energy vehicles (NEV) as long-term plans for the automotive industry (BASF, 2020).

*Governmental Regulations and Support of Green Development*

One of the important methods to prevent emissions from automobiles is applying stricter regulations for the automotive industry. The new China VI emission standards aim to reduce emissions of nitrogen oxides and other pollutants 40% to 50% by 2023 (Tabeta, 2019). They are tougher than the Euro VI standard that is in effect in Europe.

Besides, in order to enhance the BEV sales, the Chinese government has implemented various measures, such as tax exemptions, subsidies for new energy vehicles purchasing and many financial subventions for Chinese automotive makers on R&D for new technologies and production costs reduction.

Since 2014, the volume of BEVs production has been increasing constantly in the Chinese market. In comparison with other automotive markets, around more than three million electric vehicles were estimated to be sold in China while there were just under 500,000 units in the German market (Wagner, 2020a).

## Environmental Performance in the Supply Chain

In order to reduce the emitted pollutants to the environment, automotive companies should not only pay attention with the energetic operation of the BEV but also its life cycle in the whole supply chain from resource extraction, to manufacturing, usage and finally recycling. They also affect the entire environmental performance of enterprises.

Chinese automotive companies have tried to apply environmental practices in their whole supply chain such as the ISO 14001 certification and cleaner production systems. Being known as China's leading automaker, Geely seeks for sustainability development as its prioritised strategy (Tan, 2020). With not only focusing on developing and producing pure-electric vehicles, the company has also introduced green concepts and energy saving technologies in its factories. Additionally, using recyclable material is encouraged in the supply chain of Geely's vehicle production. The recoverability rate of Geely's vehicle materials has reached 96.8% and the recyclability rate is at 94.4% (Tan, 2020).

Suppliers are an important factor impacting to the environmental performance in the supply chain management. Therefore, Geely has required strictly from its suppliers to pass environmental certifications such as ISO 14001 and OHSAS 18000. It pushes its suppliers to use eco-friendly products as much as possible (Geely Holding Group, 2020).

Though having awareness with GSCM practices, the Chinese automotive supply chains are still behind compared to other developed countries.

Due to the immature consumer automotive market in China, the investment in recovery and recycling of used cars and materials are not under attention. Most of the automotive enterprises in China are still in the consideration to implement GSCM and it has therefore not yet resulted in an improved economic performance (Zhu et al., 2007).

### 4.5.5.2 Japanese Automotive Market—Toyota's SSCM

As the automotive leader in the Asian market, Japan is aware of environmental problems from automotive industry growth. Therefore, since the late 1990s, the Japanese government has encouraged automotive companies to seek for emissions reductions, reuse and recycling techniques in their manufacturing (Zhu et al., 2010). In order to gain a higher competitive position in the EU market, Japanese manufacturers have applied GSCM to fulfil the RoHS directive of the European Union (King et al., 2005).

In the domestic market, social responsibility and environmental image are the most effecting values of automotive brands. Without paying attention to environment and society, automotive companies will lose market share. Therefore, most OEMs have started producing green mobility vehicles which show less emissions to the environment. Despite being the leading automotive nation in the world, Japanese OEMs are lagging behind other global rivals in the battery electric vehicles pillar (Westbrook, 2019). Since the first mass-produced hybrid vehicle, the Toyota Prius, successfully launched in the market in 1997, Japanese OEMs have not invested much in the pure-electric cars production. Meanwhile other OEMs in China and Europe have affirmed that electric cars will be the new future of the automotive industry (Westbrook, 2019).

To catch up with other global competitors, Japanese giant companies have created joint ventures to invest in electric vehicles. For example, Toyota and Panasonic will establish one company together which will produce electric batteries used for electric vehicles by the end of 2020 (The Japan Times, 2019). With the joint venture model, Japan wants to accelerate the growth of electric vehicles and enhance their price competitiveness to compete with rivals from China and Europe. With the efforts on EVs investments, the number of registered EVs has been rising in Japan. In 2018, it reached almost 27,000 new cars (Engelmann, 2020).

*Sustainable Goals of Toyota*

In order to overcome environmental issues, Toyota is working with six sustainable development goals to be achieved for its business towards 2050 (Toyota Motor Corporation, 2019):

- Reducing $CO_2$ emissions from the vehicle's operation by 90%
- Eliminating $CO_2$ emissions during the entire vehicle life cycle
- Targeting zero emissions for all factory plants by 2050
- Minimising and optimising the water usage
- Investing more in recycling technologies and systems for End-of-life vehicle treatment
- Connecting future society with nature conservation activities

*Environmental Action Plans*

To achieve these sustainable goals, Toyota has deployed action plans. Toyota believes that electric eco-friendly vehicles are the solution to reach the goal of a $CO_2$ emissions reduction per vehicle by 90% by 2050 (The Japan Times, 2019). Therefore, the company has accelerated developing technologies for hybrid electric vehicles (HEV), plug-in hybrid electric vehicles (PHEVs), battery electric vehicles (BEVs) and fuel cell electric vehicles (FCEVs). As the leader among Japanese OEMs, Toyota has set up its target on EVs sales up to 5.5 million units in 2030 (The Japan Times, 2019).

Some electric vehicles are produced with high $CO_2$ emissions in their manufacturing process. Therefore, to reduce environmental impacts of these vehicles, it is necessary to prevent emissions throughout the entire vehicle life cycle from material parts manufacturing and vehicle assembly to distributions (Toyota Motor Corporation, 2019). Toyota has introduced the Eco-Vehicle Assessment System (Eco-VAS) which will estimate the environmental performance of the vehicle life cycle at all stages. In addition, the company also uses lightweight parts and eco-friendly materials and improves technologies for fuel efficiency, everything with the goal to reduce the $CO_2$ emissions. Logistical activities such as transportation of material, parts and completed vehicles between suppliers and to customers foster $CO_2$ emissions too. Therefore, usage of efficient fuels, shortening of logistics routes and eco-friendly transportation models are also implemented. Moreover, Toyota is seeking for zero $CO_2$ emissions in its plants. The company has introduced low-emitted $CO_2$ technologies and infrastructures with renewable energies such as solar power and wind power along with hydrogen energy (Toyota Motor Corporation, 2019).

## References

Abdul-Muhmin, A. G. (2007). Explaining consumers' willingness to be environmentally friendly. *International Journal of Consumer Studies, 31*(3), 237–247. https://doi.org/10.1111/j.1470-6431.2006.00528.x

Abrenica, J. V. (1998). The Asian automotive industry: Assessing the roles of state and market in the age of global competition. *Asian-Pacific Economic Literature, 12*(1), 12–26. https://doi.org/10.1111/1467-8411.00026

Autry, C. W. (2005). Formalization of reverse logistics programs: A strategy for managing liberalized returns. *Industrial Marketing Management, 34*(7), 749–757. https://doi.org/10.1016/j.indmarman.2004.12.005

Balon, V., Sharma, A. K., & Barua, M. K. (2016). Assessment of barriers in green supply chain management using ISM: A case study of the automobile industry in India. *Global Business Review, 17*(1), 116–135. https://doi.org/10.1177/0972150915610701

BASF. (2020). *Challenges and opportunities for China's automotive market*. https://www.basf.com/cn/en/media/BASF-Information/Resources-environment-climate/Challenges-and-opportunities-for-China-automotive-market.html

BMW Group. (2018). *Environmental statement BMW group 2018: Environmental protection in production*. https://www.bmwgroup.com/content/dam/grpw/websites/bmwgroup_com/responsibility/downloads/en/2018/2018-BMW-Group-Environmental-Statement.pdf

BMW Group. (2020). *Sustainable value report 2019*. https://www.bmwgroup.com/content/dam/grpw/websites/bmwgroup_com/responsibility/downloads/de/2020/2020-BMW-Group-SVR-2019-Deutsch.pdf

Borade, A. B., & Bansod, S. V. (2007). Domain of supply chain management: A state of art. *Journal of Technology Management & Innovation, 2*(4), 109–121.

Bormann, R., Fink, P., & Holzapfel, H. (2018). *The future of the German automotive industry: Transformation by disaster or by design. WISO-Diskurs: 10/2018*. Friedrich-Ebert-Stiftung, Division of Economic and Social Policy.

Bundesministerium für Bildung und Forschung. (2020). *Open Hybrid LabFactory*. https://www.forschungscampus.bmbf.de/forschungscampi/ohlf

Carbon Disclosure Project. (2011). *Carbon Disclosure Project: Supply chain report 2011*. https://www.marriott.com/marriottassets/Multimedia/PDF/CorporateResponsibility/serve360/CDP-2011-Supply-Chain-Report.pdf

Chan, F. T. S., & Kumar, N. (2007). Global supplier development considering risk factors using fuzzy extended AHP-based approach. *Omega, 35*(4), 417–431. https://doi.org/10.1016/j.omega.2005.08.004

Chandramowli, S., Transue, M., & Felder, F. A. (2011). Analysis of barriers to development in landfill communities using interpretive structural modeling. *Habitat International, 35*(2), 246–253. https://doi.org/10.1016/j.habitatint.2010.09.005

Chen, Y. J., & Sheu, J. B. (2009). Environmental-regulation pricing strategies for green supply chain management. *Transportation Research Part E: Logistics and Transportation Review, 45*(5), 667–677.

Commission for Environmental Cooperation. (2005). *Successful practices of environmental management systems in small and medium-sized enterprises: A North American perspective*. Commission for Environmental Cooperation. http://www3.cec.org/islandora/en/item/2273-successful-practices-environmental-management-systems-in-small-and-medium-size-en.pdf

Cooper-Searle, S. (2017). *Making climate friendly cars: Material considerations*. https://hoffmanncentre.chathamhouse.org/article/reusing-and-recycling-car-materials/

Crainic, T. G., Gendreau, M., & Dejax, P. (1993). Dynamic and stochastic models for the allocation of empty containers. *Operations Research, 41*(1), 102–126. https://doi.org/10.1287/opre.41.1.102

Daimler AG. (2020). *Sustainability report 2019*. https://www.daimler.com/documents/sustainability/other/daimler-sustainability-report-2019.pdf

Dekker, R., Bloemhof, J., & Mallidis, I. (2012). Operations research for green logistics: An overview of aspects, issues, contributions and challenges. *European Journal of Operational Research, 219*(3), 671–679. https://doi.org/10.1016/j.ejor.2011.11.010

Diabat, A., & Govindan, K. (2011). An analysis of the drivers affecting the implementation of green supply chain management. *Resources, Conservation and Recycling, 55*(6), 659–667. https://doi.org/10.1016/j.resconrec.2010.12.002

EDF Energy. (2020). *All about electric car batteries.* https://www.edfenergy.com/electric-cars/batteries

Emmett, S., & Sood, V. (2010). *Green supply chains: An action manifesto.* John Wiley & Sons Inc..

Engelmann, J. (2020). *Electric vehicles in Japan - Statistic & facts.* https://www.statista.com/topics/5628/electric-vehicles-in-japan/

Figenbaum, A., & Thomas, H. (1986). Dynamic and risk measurement perspectives on bowman's risk-return paradox for strategic management: An empirical study. *Strategic Management Journal, 7*(5), 395–407. https://doi.org/10.1002/smj.4250070502

Gaudillat, P. F., Antonopoulos, I. S., Dri, M., Canfora, P., & Traverso, M. (2017). *Best environmental management practice for the car manufacturing sector: Learning from frontrunners. JRC science for policy report.* Publications Office of the European Union. https://doi.org/10.2760/202143. https://publications.europa.eu/en/publication-detail/-/publication/f742641a-f9a6-11e7-b8f5-01aa75ed71a1/language-en/format-PDF/source-65246120

Gavronski, I., Klassen, R. D., Vachon, S., & Nascimento, L. F. M. D. (2011). A resource-based view of green supply management. *Transportation Research Part E: Logistics and Transportation Review, 47*(6), 872–885. https://doi.org/10.1016/j.tre.2011.05.018

Geely Holding Group. (2020). *Corporate social responsibility report: 2019.* http://geelyauto.com.hk/core/files/corporate_governance/en/2019_CORPORATE_SOCIAL_RESPONSIBILITY_REPORT.pdf

Germany Trade & Invest. (2015). *Electromobility in Germany: Vision 2020 and beyond.* http://v2city-expertgroup.eu/wp-content/uploads/2016/02/electromobility-in-germany-vision-2020-and-beyond-en.pdf

Germany Trade & Invest. (2018). *The automotive industry in Germany.* https://www.gtai.de/resource/blob/64100/817a53ea3398a88b83173d5b800123f9/industry-overview-automotive-industry-en-data.pdf

Gifford, D. J. (1997). The value of going green. *Harvard Business Review, 75*(5), 11–12.

Golicic, S. L., & Smith, C. D. (2013). A meta-analysis of environmentally sustainable supply chain management practices and firm performance. *Journal of Supply Chain Management, 49*(2), 78–95. https://doi.org/10.1111/jscm.12006

Götze, S. (2019). *The downside of electromobility: Lithium mining in South America destroys livelihoods & access to water for indigenous people.* Business & Human Rights Resource Centre. https://www.business-humanrights.org/en/the-downside-of-electromobility-lithium-mining-in-south-america-destroys-livelihoods-access-to-water-for-indigenous-people

Green Purchasing Guide. (2011). *Commitment to buy green. Greening greater Toronto.* http://www.partnersinprojectgreen.com/files/GreenPurchasing Guide.pdf

Grote, C. A., Jones, R. M., Blount, G. N., Goodyer, J., & Shayler, M. (2007). An approach to the EuP directive and the application of the economic eco-design for complex products. *International Journal of Production Research, 45*(18–19), 4099–4117. https://doi.org/10.1080/00207540701450088

Hawks, K. (2006). What is reverse logistics? *Reverse Logistics Magazine,* Winter/Spring.

Hotten, R. (2015). *Volkswagen: The scandal explained.* https://www.bbc.com/news/business-34324772

Hsu, C.-W., & Hu, A. H. (2008). Green supply chain management in the electronic industry. *International Journal of Environmental Science & Technology, 5*(2), 205–216. https://doi.org/10.1007/BF03326014

Hunke, K., & Prause, G. (2014). Sustainable supply chain management in German automotive industry: Experiences and success factors. *Journal of Security and Sustainability Issues, 3*(3), 15–22. https://doi.org/10.9770/jssi.2014.3.3(2)

Igarashi, M., de Boer, L., & Fet, A. M. (2013). What is required for greener supplier selection?: A literature review and conceptual model development. *Journal of Purchasing and Supply Management, 19*(4), 247–263.

ISO. (2015). *ISO 14000 family – Environmental management.* https://www.iso.org/iso-14001-environmental-management.html

Jenu, S., Deviatkin, I., Hentunen, A., Myllysilta, M., Viik, S., & Pihlatie, M. (2020). Reducing the climate change impacts of lithium-ion batteries by their cautious management through integration of stress factors and life cycle assessment. *Journal of Energy Storage, 27,* 101023. https://doi.org/10.1016/j.est.2019.101023

Kallstrom, H. (2019). *Suppliers' power is increasing in the automobile industry.* https://marketrealist.com/2015/02/suppliers-power-increasing-automobile-industry/

Kannan, G., Pokharel, S., & Kumar, P. S. (2009). A hybrid approach using ISM and fuzzy TOPSIS for the selection of reverse logistics provider. *Resources,*

*Conservation and Recycling, 54*(1), 28–36. https://doi.org/10.1016/j.resconrec.2009.06.004

King, A. A., Lenox, M. J., & Terlaak, A. (2005). The strategic use of decentralized institutions: Exploring certification with the ISO 14001 management standard. *Academy of Management Journal, 48*(6), 1091–1106. https://doi.org/10.5465/amj.2005.19573111

Kumar, A., Jain, V., & Kumar, S. (2014). A comprehensive environment friendly approach for supplier selection. *Omega, 42*(1), 109–123. https://doi.org/10.1016/j.omega.2013.04.003

Kuo, T. C., Ma, H.-Y., Huang, S. H., Hu, A. H., & Huang, C. S. (2010). Barrier analysis for product service system using interpretive structural model. *The International Journal of Advanced Manufacturing Technology, 49*(1–4), 407–417. https://doi.org/10.1007/s00170-009-2399-7

Lakshmimeera, B. L., & Palanisamy, C. (2013). A conceptual framework on green supply chain management practices. *Industrial Engineering Letters, 3*(10), 42–52.

Li, C., Liu, F., & Wang, Q. (2010). Planning and implementing the green manufacturing strategy: Evidences from western China. *Journal of Science and Technology Policy in China, 1*(2), 148–162. https://doi.org/10.1108/17585521011059884

McGee, P. (2018). *What went so right with Volkswagen's restructuring?* https://www.ft.com/content/a12ec7e2-fa01-11e7-9b32-d7d59aace167

McKinnon, A. C. (2005). The economic and environmental benefits of increasing maximum truck weight: The British experience. *Transportation Research Part D: Transport and Environment, 10*(1), 77–95.

McKinsey. (2020). *The road to 2020 and beyond: Whats driving the global automotive industry*. https://www.mckinsey.com/industries/automotive-and-assembly/our-insights/the-road-to-2020-and-beyond-whats-driving-the-global-automotive-industry

Michel, V., & Siegfried, P. (2021). Digitale Speditionen in der Lebensmittellogistik digital freight forwarders in food logistics. *Logistics Journal*. https://doi.org/10.2195/lj_NotRev_michel_de_202102_01. ISSN 1860-5923.

Min, H., & Galle, W. P. (1997). Green purchasing strategies: Trends and implications. *International Journal of Purchasing and Materials Management, 33*(2), 10–17. https://doi.org/10.1111/j.1745-493X.1997.tb00026.x

Moore, M. (2020). *Number of passenger cars sold in the Asia Pacific region 2019, by country or region*. https://www.statista.com/statistics/584904/asia-pacific-passenger-car-sales-by-country/

Muduli, K., Govindan, K., Barve, A., & Geng, Y. (2013). Barriers to green supply chain management in Indian mining industries: A graph theoretic approach. *Journal of Cleaner Production, 47*, 335–344. https://doi.org/10.1016/j.jclepro.2012.10.030

Nunes, B., & Bennett, D. (2010). Green operations initiatives in the automotive industry. *Benchmarking: An International Journal, 17*(3), 396–420. https://doi.org/10.1108/14635771011049362

Office of Energy Efficiency & Renewable Energy. (2020). *Lightweight materials for cars and trucks.*

Olaore, R. (2013). Accounting, purchasing and supply chain management Interface. *IOSR Journal of Business and Management, 11*(2), 80–84. https://doi.org/10.9790/487X-1128084

Oumer, A. J., Cheng, J. K., & Tahar, R. M. (2015). *Green manufacturing and logistics in automotive industry: A simulation model.* 9th International Conference on IT in Asia (CITA), 1–6. https://doi.org/10.1109/CITA.2015.7349839.

Pereseina, V., Jensen, L.-M., Hertz, S., & Cui, L. (2014). Challenges and conflicts in sustainable supply chain management: Evidence from the heavy vehicle industry. *Supply Chain Forum: An International Journal, 15*(1), 22–32. https://doi.org/10.1080/16258312.2014.11517331

Petroff, A. (2018). *Carmakers and big tech struggle to keep batteries free from child labor.* https://money.cnn.com/2018/05/01/technology/cobalt-congo-child-labor-car-smartphone-batteries/index.html

PricewaterhouseCoopers. (2014). *How to be No. 1: Facing future challenges in the automotive industry.* https://www.pwc.com.tr/tr/publications/industrial/automotive/pdf/otomotiv-sektorunu-bekleyen-zorluklari-asmak.pdf

Punit, S., Yash, R., Shridhar, S., & Rohit, Y. (2015). A review on green supply chain management in automobile industry. *International Journal of Current Engineering and Technology, 5*(6), 3697–3702. http://inpressco.com/category/ijcet

Rogers, D. S., & Tibben-Lembke, R. S. (1999). *Going backwards: Reverse logistics trends and practices.* Reverse Logistics Executive Council.

Sanket Tonape, M. O. (2013). An overview, trends and future mapping of green supply chain management – Perspectives in India. *Journal of Supply Chain Management Systems, 2*(3).

Sarkis, J. (2006). *Greening the supply chain.* Springer.

Schwartz, J. (2018). *VW investors sue for billions of dollars over diesel scandal.* https://www.reuters.com/article/us-volkswagen-emissions-trial/vw-investors-sue-for-billions-of-dollars-over-diesel-scandal-idUSKCN1LQ0W4

Seuring, S., & Müller, M. (2008). From a literature review to a conceptual framework for sustainable supply chain management. *Journal of Cleaner Production, 16*(15), 1699–1710. https://doi.org/10.1016/j.jclepro.2008.04.020

Shao, J., & Ünal, E. (2019). What do consumers value more in green purchasing?: Assessing the sustainability practices from demand side of business. *Journal of Cleaner Production, 209*, 1473–1483. https://doi.org/10.1016/j.jclepro.2018.11.022

Siegfried, P. (2021). *Business management case studies*, ISBN: 978-3-75431-691-7. BoD Book on Demand.

Siegfried, P., Michel, A., Tänzler, J., & Zhang, J. (2021). Analysing sustainability issues in urban logistics in the context of growth of e-commerce. *Journal of Social Sciences, IV*(1), 6–11. ISSN: 2587-3490.

Singh, A. (2020). *Electric vehicle market size, share, analysis, growth by 2027.* https://www.alliedmarketresearch.com/electric-vehicle-market

Srivastava, S. K. (2007). Green supply-chain management: A state-of-the-art literature review. *International Journal of Management Reviews, 9*(1), 53–80. https://doi.org/10.1111/j.1468-2370.2007.00202.x

Sroufe, R. (2003). Effects of environmental management systems on environmental management practices and operations. *Production and Operations Management, 12*(3), 416–431. https://doi.org/10.1111/j.1937-5956.2003.tb00212.x

Sustainable Business Toolkit. (2015). *11 ways green distribution can be sustainable.* https://www.sustainablebusinesstoolkit.com/green-distribution/

Svensson, G. (2007). Aspects of sustainable supply chain management (SSCM): Conceptual framework and empirical example. *Supply Chain Management: An International Journal, 12*(4), 262–266. https://doi.org/10.1108/13598540710759781

Tabeta, S. (2019). China's new emissions rules take scalpel to bloated auto industry. *Nikkei Asian Review.* https://asia.nikkei.com/Business/Business-trends/China-s-new-emissions-rules-take-scalpel-to-bloated-auto-industry

Tan, D. (2020). *Geely banning single-use plastics in all its facilities.* https://paultan.org/2020/04/29/geely-banning-single-use-plastics-in-all-its-facilities/

Tandem Logistics. (2020). *Glossary of terms.* https://tandemlogistics.com/glossary-of-terms/

The Japan Times. (2019). *Toyota and Panasonic to launch joint venture to make electric vehicle batteries by end of 2020.* https://www.japantimes.co.jp/news/2019/01/23/business/corporate-business/toyota-panasonic-launch-joint-venture-make-electric-vehicle-batteries-end-2020/

Thompson, M. (2018). *CNN investigation: Daimler promises to audit cobalt supply 'to the mine'*. https://money.cnn.com/2018/05/02/investing/daimler-cobalt-supply-chain/index.html

Toyota Motor Corporation. (2019). *Environmental report 2019*. https://global.toyota/pages/global_toyota/sustainability/report/er/er19_en.pdf

Verband der Automobilindustrie. (2010). *The national platform for electric mobility*. https://www.vda.de/en/topics/innovation-and-technology/electromobility/National-Platform-for-Electric-Mobility.html

Volkswagen AG. (2020a). *History*. https://www.volkswagenag.com/en/group/history.html

Volkswagen AG. (2020b). *Sustainability report 2019*. https://www.volkswagenag.com/presence/nachhaltigkeit/documents/sustainability-report/2019/Nonfinancial_Report_2019_e.pdf

Wagner, I. (2020a). *Estimated electric vehicles in use in selected countries as of 2019*. https://www.statista.com/statistics/244292/number-of-electric-vehicles-by-country/

Wagner, I. (2020b). *Worldwide automobile production through 2019*. https://www.statista.com/statistics/262747/worldwide-automobile-production-since-2000/

Walker, H., Di Sisto, L., & McBain, D. (2008). Drivers and barriers to environmental supply chain management practices: Lessons from the public and private sectors. *Journal of Purchasing and Supply Management, 14*(1), 69–85. https://doi.org/10.1016/j.pursup.2008.01.007

Westbrook, J. T. (2019). The Japanese auto industry is finally getting serious about electric cars. *Jalopnik*. https://jalopnik.com/the-japanese-auto-industry-is-finally-getting-serious-a-1839256462

Wisner, J. D., Tan, K.-C., & Leong, G. K. (2012). *Principles of supply chain management: A balanced approach* (3rd ed.). South-Western/Cengage Learning.

Wong, S. (2020). *Car sales (passenger and commercial vehicles) in China from 2009 to 2019*. https://www.statista.com/statistics/233743/vehicle-sales-in-china/

Xia, Y., & Tang, L.-P. (2011). Sustainability in supply chain management: Suggestions for the auto industry. *Management Decision, 49*(4), 495–512. https://doi.org/10.1108/00251741111126459

Zhu, Q., Geng, Y., Fujita, T., & Hashimoto, S. (2010). Green supply chain management in leading manufacturers. *Management Research Review, 33*(4), 380–392. https://doi.org/10.1108/01409171011030471

Zhu, Q., & Sarkis, J. (2004). Relationships between operational practices and performance among early adopters of green supply chain management practices in Chinese manufacturing enterprises. *Journal of Operations Management*, 265–289.

Zhu, Q., & Sarkis, J. (2006). An inter-sectoral comparison of green supply chain management in China: Drivers and practices. *Journal of Cleaner Production, 14*(5), 472–486. https://doi.org/10.1016/j.jclepro.2005.01.003

Zhu, Q., Sarkis, J., & Lai, K.-h. (2007). Green supply chain management: Pressures, practices and performance within the Chinese automobile industry. *Journal of Cleaner Production, 15*(11–12), 1041–1052. https://doi.org/10.1016/j.jclepro.2006.05.021

# 5

# Scenarios and Concepts for the Future Development

## 5.1 Partnerships Between the Manufacturers

Facing environmental changes has led the automotive industry to sustainable development. Therefore, it is necessary for automotive companies to fast respond to this trend in order to gain higher market shares. Bringing new vehicles to the market puts many challenges on OEMs. Therefore, the concept of co-development partnerships should be used between OEMs to reach a shortening of time-to-market, to increase the profitability, to enhance the innovation in R&D and to expand the market access (Chesbrough & Schwartz, 2007). These partners could be competitors, suppliers or partners (Sawhney et al., 2005; Siegfried, 2015a; Siegfried, 2015b). Therefore, to overcome barriers in the implementation of a green supply chain, German OEMs should develop partnerships to be able to maintain their competitive advantages.

As mentioned before, to overtake other global competitors in the EV production, Japanese giant companies have created joint ventures to invest in electric vehicles. Toyota and Panasonic will establish a company together which will produce batteries used for electric vehicles by the end of 2020 (The Japan Times, 2019). With the joint venture model, Japan wants to accelerate the growth of electric vehicles and enhance their price competitiveness to compete with rivals from China and Europe. The success of the partnership model in Japan can be seen as a good example for German OEMs.

By collaborating with partners, OEMs can improve their knowledge and expand innovation in terms of sharing know-how and technology in the development of a new product or service. Investments in green supply chain management lead to higher cost pressure (McKinsey, 2020). In addition, the automotive industry consists of long global supply chain networks that can cause extremely high transportation costs for companies (Xia & Tang, 2011). Hence, partnerships will create more economic advantages for OEMs by dividing costs in material sourcing, transportation, production, distribution and gathering budgets for R&D (Chesbrough & Schwartz, 2007). Cooperation and joint ventures with partners can be the strategic model for automotive companies on the path to developing a green supply chain by exchanging and pooling complementary resources and capabilities between different firms (Grant, 2008).

## 5.2 Extending Environmental Initiatives Throughout Supply Chain Actors

Although most OEMs have orientated towards implementing environmental protection initiatives into their supply chain management, it still remains the lack of top management commitments due to high costs of green investments and the slow rate of benefit returns (Balon et al., 2016). Besides, OEMs are still facing with missing knowledge among employees and workers in terms of using new green innovative technologies.

Hence, it is necessary for automotive companies to establish overall directions and useful guidelines for all stakeholders and actors along the supply chain in the next years. They need to find an effective way to transfer knowledge and develop stakeholders' awareness on benefits and advancements achieved by implementing green supply chain management. Through best practices and case studies of succeeding companies in SSCM, through training courses and capacity building programmes, all of the actors along the supply chain can be motivated to pursue sustainable concepts. Moreover, OEMs will also get benefits from these environmental trainings. The employees and workers with better know-how in the new green technologies will help to enhance the firm's environmental and economic performance.

## 5.3 Customer Orientation as a High Influence Factor

The results and findings from the customer survey (see Appendix) point out that automotive customers are becoming more aware of sustainable concept within the entire lifecycle of vehicles that can reduce negative effects on the environment. Compared to the past, the interest on green vehicles has increased which has been proven by growing EV sales (Virta, 2020).

This survey results also show that customers are not only paying attention to their vehicle's emissions and energy efficiency during the operation but also to the entire environmental activities of the whole value chain needed to produce it. This includes the questions whether materials or components are made from renewable sources, whether component suppliers are sustainable, whether a sustainable vehicle assembly process is used, whether sustainable ways of components transportation and vehicle distribution are used and whether vehicles and its components can be recycled after their end-of-life. Presenting the responses of the survey's participants to these questions, Fig. 5.1 shows customers' overall high importance grade of these issues for their vehicle buying decision.

Customers are the most important factor affecting directly to the business' revenues and benefits of a company. The more customers can be gained, the more revenues can be reached (Apte & Sheth, 2017). According to the survey, more than 90% of the participants show higher trust in companies that implement SSCM. The majority of them also accept paying higher prices for vehicles which are produced using SSCM. That results in higher revenues and creates more values for OEMs.

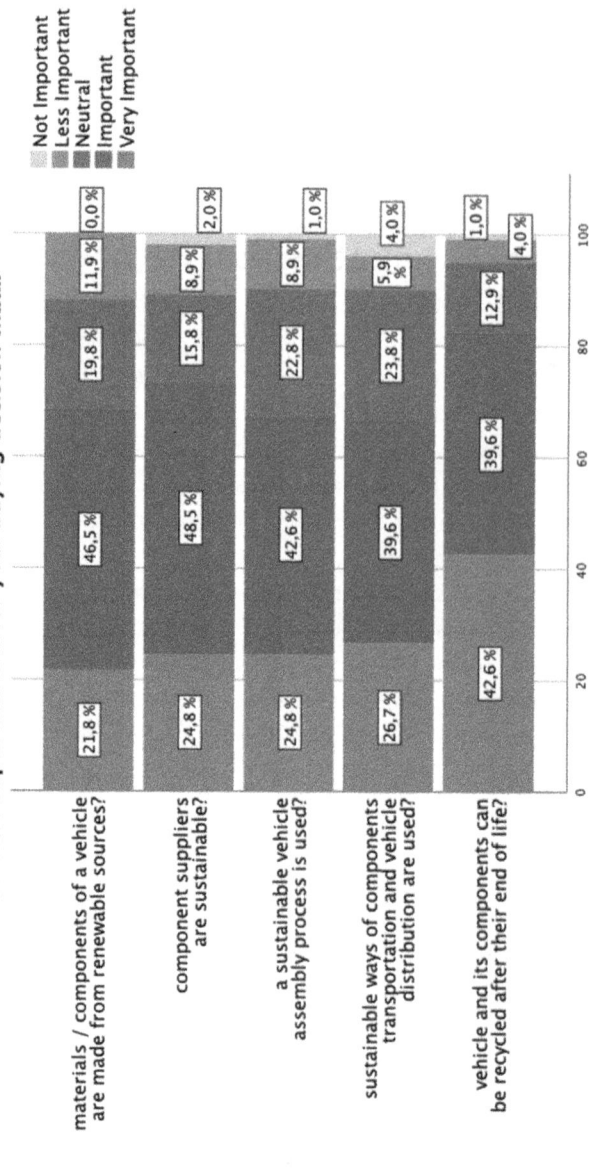

Fig. 5.1 Importance of buying decisions

Therefore, OEMs need to orientate their strategies towards customer expectations and their values and opinions should impact the companies' decision-making.

## 5.4 Providing Information to Customers Through SSC Labelling

In the current automotive market, there is an energy-efficiency label giving details about fuel efficiency and $CO_2$ emissions to be attached to all new cars. This label is rated in classes from A+++ to D which helps drivers to choose cars with a low energy consumption. However, it still lacks the information about sustainable standards of the vehicle's whole supply chain activities (e.g. material and supplier origins, compliance with environmental regulations in the assembly, distribution and recycling processes). In this research's survey (see appendix) more than 50% of participants confirmed that they have difficulties of finding information about how sustainable the supply chain of current vehicles is (see Fig. 5.2).

To improve this in the future, a possible suggestion to manufactures is to invent a sustainable supply chain (SSC) label which contains details about the vehicle's entire life cycle. It should include all processes from material extraction to recycling and ensure that they all have been operated with environmental protection initiatives and all actors in the supply chain have proven their environmental certifications. In the survey it was also shown that the majority of customers are willing to pay higher price for the SSC labelled vehicle or a vehicle with a higher rating compared to competitors (see Fig. 5.3).

This finding reveals that a mandatory SCM label brings economic advantages to companies. The higher the SSC label is rated, the better the environmental image build-up of the OEM can be which can have a positive influence on the customers' loyalty and retention. This creates long-term economic benefits through higher sales and higher revenues. Another possible positive effect of such label is the pressure it puts on a company's competitor to achieve similar ratings and therefore also to invest more in SSCM.

19. How easy is it for you to find information about the sustainability of the supply chain of vehicles in the current market (e.g. material and supplier origins, compliance with environmental regulations in the assembly, distribution and recycling processes)?

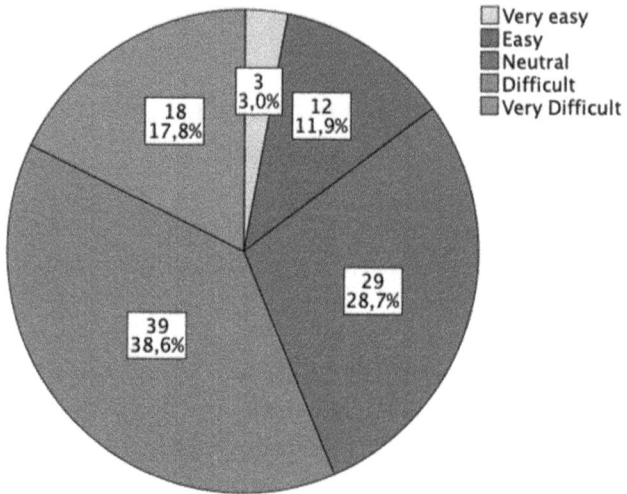

**Fig. 5.2** Finding information about sustainability

20. How likely will you accept to pay a higher price for a vehicle with a higher rated SSC label compared to a vehicle with a lower rating or even no label at all?

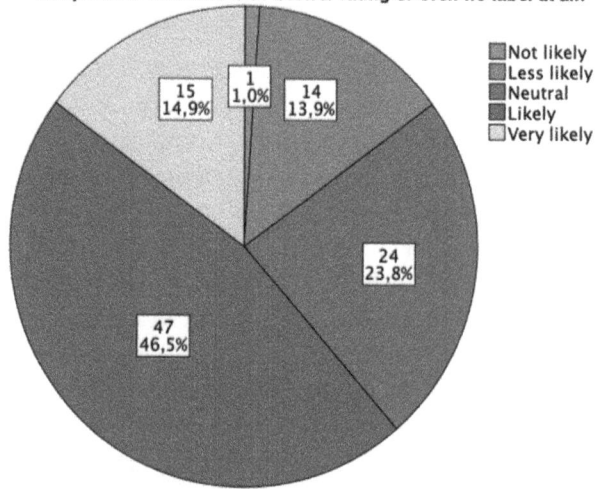

**Fig. 5.3** Accepting higher prices

# References

Apte, S., & Sheth, J. (2017). Developing the sustainable edge. *Leader to Leader, 2017*(85), 48–53. https://doi.org/10.1002/ltl.20306

Balon, V., Sharma, A. K., & Barua, M. K. (2016). Assessment of barriers in green supply chain management using ISM: A case study of the automobile industry in India. *Global Business Review, 17*(1), 116–135. https://doi.org/10.1177/0972150915610701

Chesbrough, H., & Schwartz, K. (2007). Innovating business models with co-development partnerships. *Research-Technology Management, 50*(1), 55–59. https://doi.org/10.1080/08956308.2007.11657419

Grant, R. M. (2008). *Contemporary strategy analysis* (6th ed.). Blackwell.

McKinsey. (2020). *The road to 2020 and beyond: Whats driving the global automotive industry.* https://www.mckinsey.com/industries/automotive-and-assembly/our-insights/the-road-to-2020-and-beyond-whats-driving-the-global-automotive-industry

Sawhney, M., Verona, G., & Prandelli, E. (2005). Collaborating to create: The internet as a platform for customer engagement in product innovation. *Journal of Interactive Marketing, 19*(4), 4–17. https://doi.org/10.1002/dir.20046

Siegfried, P. (2015a). *Business cases internationalisation strategies in global player companies*: Volume 1 - ISBN: 978-3-86924-614-7. AVM Akademische Verlagsgemeinschaft.

Siegfried, P. (2015b). *Business cases internationalisation strategies in global player companies*: Volume 2 - ISBN: 978-3-86924-628-4. AVM Akademische Verlagsgemeinschaft.

The Japan Times. (2019). *Toyota and Panasonic to launch joint venture to make electric vehicle batteries by end of 2020.* https://www.japantimes.co.jp/news/2019/01/23/business/corporate-business/toyota-panasonic-launch-joint-venture-make-electric-vehicle-batteries-end-2020/

Virta. (2020). *The global electric vehicle market in 2020: Statistics & forecasts.* https://www.virta.global/global-electric-vehicle-market

Xia, Y., & Tang, L.-P. (2011). Sustainability in supply chain management: Suggestions for the auto industry. *Management Decision, 49*(4), 495–512. https://doi.org/10.1108/00251741111126459

# 6

# Conclusion of Sustainable Supply Chain Management: Learning from the German Automotive Industry

The automotive sector is considered to be one of the fastest developing industries in the world and contributes to economic growth in many countries. This development however results in an increase of $CO_2$ emissions based on higher numbers of vehicles driving on the roads. Besides, this industry causes other negative impacts on the environment by the supply chain's activities like material extracting, burning fossil fuels in the production plants and releasing waste. In the recent years, people are paying more attention to environmental issues. Since this topic becomes more urgent, many environmental protection activities are suggested to minimise the harmful influences from the automotive industry. To find long-term solutions, automotive manufacturers have researched and integrated the sustainability concept into their entire supply chain management. The term sustainable supply chain management is, therefore, recognised as an important strategic development concept for automotive companies.

This research investigated the important values of practising SSCM for manufacturers through environmental and economic aspects. Major driving factors leading to the implementation of SSCM were identified as competitiveness, governmental regulations and customers' buying behaviour. The general research objectives in this research were to analyse current effects and challenges of implementing SSCM as well as GSCM strategies in the automotive industry, particularly in the German market. To achieve the research objectives, a literature review on SSCM concepts regarding environmental and economic performances was conducted. Furthermore, green activities of companies like green material sourcing, green supplier selection, green transportation, green product design, green manufacturing, green distribution and green recycling systems throughout the sustainable supply chain were presented.

In addition, this study researched current successful achievements and strategies in SSCM through performing case studies on three leading German automotive manufacturers, VW, BMW and Daimler. It could be seen that these companies have implemented the sustainable initiatives in their supply chain processes in effective ways. For instance, all of them have obtained environmental certifications like ISO 14001 and EMAS. Besides targeting to produce more sustainable mobility solutions like electric vehicles, these companies have striven to invest and research more in green technologies for the entire supply chain to become carbon neutral.

Moreover, to understand more about their competitors and to examine the competitive advantages of German automotive manufacturers, this study also benchmarked the implementation of SSCM in other markets like China and Japan. Findings from case studies on these competitors lead to the suggestion for the German companies to collaborate more with partners and suppliers to improve their knowledge and expand innovations in terms of sharing know-how and technology in the development of green concepts.

The study also concentrated on showing barriers and challenges that OEMs are still facing with the SSCM implementation. The cost concerns might be the most serious barrier for considering environmental factors in the automotive processes. Investments in green technologies result in higher costs (Walker et al., 2008). Although many automotive companies want to practise GSCM, they are facing the issue of finding a balance between being environmentally friendly and protecting the nature, on the one hand, and meeting shareholder satisfaction with high profits on the other side (Gifford, 1997). However, establishing environmental practices in the supply chain even leads to improvements of economic benefits for OEMs. Companies using SSCM can generate more turnovers and revenues from customer sales due to their green images and reputations (Hunke & Prause, 2014).

Furthermore, this research determines the essential role of customers in influencing firms to undertake the SSCM approach. A study with over 100 potential vehicle customers using a questionnaire was performed. The results show that drivers increased their awareness and appreciation of the sustainable supply chain of the entire vehicle lifecycle. "Green" oriented customers are willing to pay a higher price for vehicles that were developed and assembled along SSCM. This information can be used by OEMs for their financial planning. Investments in SSCM could be paid back by higher profits from higher vehicle sale volumes. Practising SSCM helps companies to enhance their green images and reliable reputation in customers' eyes. Creating trust of customers could lead to a higher loyalty to their brands and therefore in continuous reoccurring sales.

Finally, this research points out suggestions for automotive OEMs' SSCM development for the next years. To develop SSCM more effectively and prevent barriers, it is necessary for automotive companies to establish overall directions and useful guidelines for all stakeholders and actors along the supply chain. Training courses and capacity building programmes help employees and managers to recognise the high benefits of approaching sustainability concepts. In possible future scenarios for the automotive industry, customers will still play important roles for the companies that affect directly to the business' revenues and benefits. Therefore, automotive manufacturers need to base their sustainable strategies depending on customer's expectations and requirements. From the opinions of the customers participating in the survey (see Appendix), it is recommended to automotive companies to invent a sustainable supply chain label and place it on their new vehicles. This label could provide more details about the sustainable activities during the entire value chain of the vehicle to possible buyers. That can build up more customer loyalty and better brand images. Thus, it could result in long-term economic benefits for automotive manufacturers through higher sales and revenues.

# References

Gifford, D. J. (1997). The value of going green. *Harvard Business Review, 75*(5), 11–12.

Hunke, K., & Prause, G. (2014). Sustainable supply chain management in German automotive industry: Experiences and success factors. *Journal of Security and Sustainability Issues, 3*(3), 15–22. https://doi.org/10.9770/jssi.2014.3.3(2)

Walker, H., Di Sisto, L., & McBain, D. (2008). Drivers and barriers to environmental supply chain management practices: Lessons from the public and private sectors. *Journal of Purchasing and Supply Management, 14*(1), 69–85. https://doi.org/10.1016/j.pursup.2008.01.007

# Appendix

## Results of Survey on Potential Automotive Customers

To find out more about the consumers' view on sustainable supply chains in the automotive industry, a customer survey was performed during the creation of this research. It should help to answer the questions whether the consumers show a higher interest in products with green supply chain characteristics, whether they are aware of the origins of the materials, whether they are renewable or recyclable and whether the availability of suppliers with green performances is even taken into consumers' considerations. A questionnaire was sent out and collected 101 usable responses. The following paragraphs give an overview about the questions that were asked, the given responses as well as their meaning for the importance of SSCM in the automotive industry.

During the distribution of the survey the aim was to reach both male and female from different age groups. As shown in Fig. A.1, among the 101 participants in the survey there are 57.4% male and 42.6% female. So, both genders are almost equally represented.

# Appendix

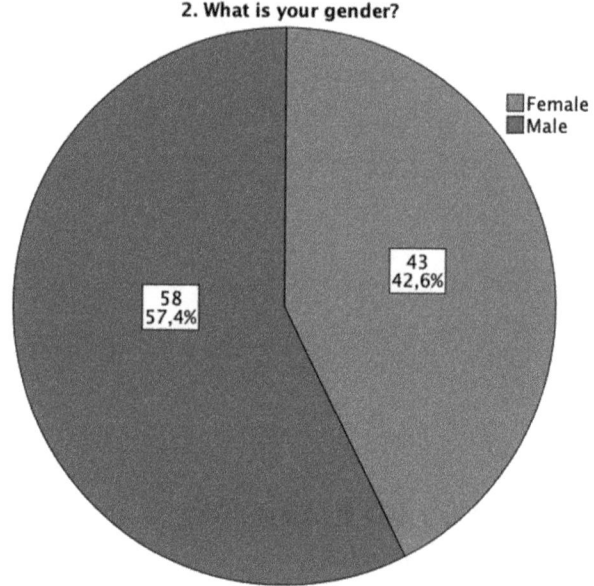

**Fig. A.1** Participants' gender

Figure A.2 shows the participating age groups. The age groups in the survey are slipped in seven ranges from under 18 (0%), 18–24 (20.8%), 25–34 (53.5%), 35–44 (9.9%), 45–54 (9.9%), 55–64 (5.9%) to over 65 years old (0%). The result shows that ages from 18 to 64 are included with a majority in the group from 25 to 34 years.

Moreover, the survey also asked the participants about other background details like their education level and employment status (Figs. A.3 and A.4).

To receive a first overview about the participants mobility usage and their attitude towards environmentally friendly vehicles they were asked, what engine their current vehicle has and how important it is for them that their vehicle emits a low amount of $CO_2$ emissions to the environment.

The results shown in Fig. A.5 expose that the majority of the participant own a vehicle with a combustion engine. Just two customers stated to own an electric vehicle. Hybrid of hydrogen vehicle drivers was not found in this sample group. The next question however shows that nearly

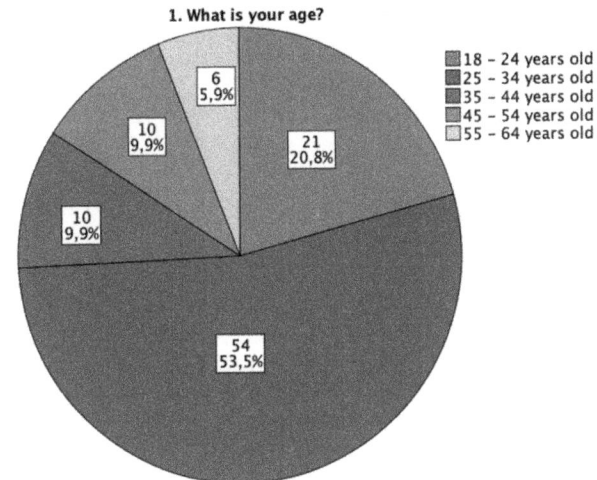

**Fig. A.2** Participants' age

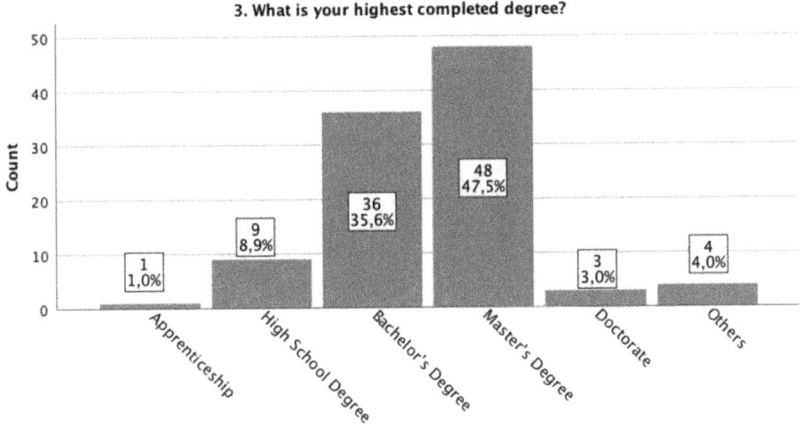

**Fig. A.3** Participants' highest completed degree

80% of participants indicated that it is important for them that their vehicles should emit less $CO_2$ emissions and toxic pollutants to the environment during their usage (Fig. A.6).

Despite this high number of participants which agreed to the importance of reduced emissions of their vehicles, just 59% of the sample group

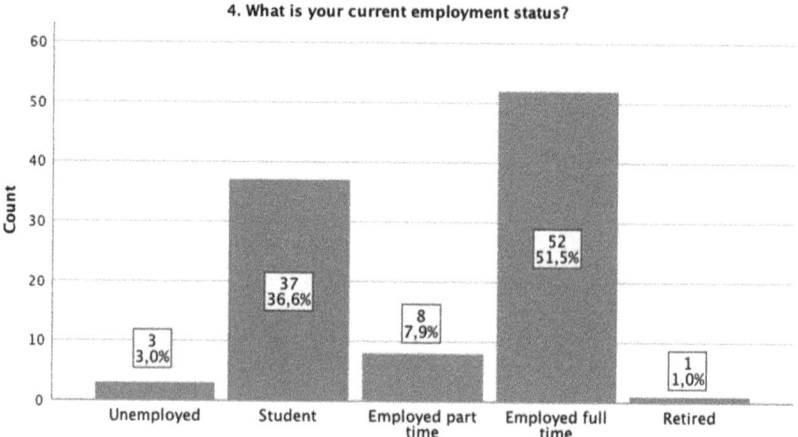

**Fig. A.4** Participants' employment status

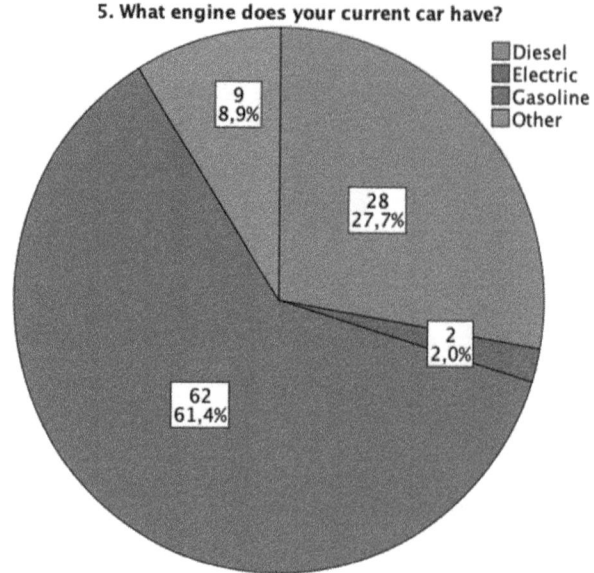

**Fig. A.5** Participants' current car engine

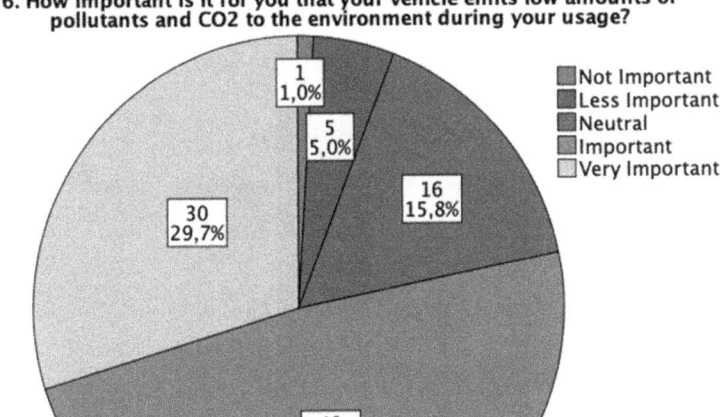

**Fig. A.6** Results of question six

stated that they want to purchase a green vehicle like an electric or hybrid electric vehicle in the future. In Fig. A.7, it can be seen that mostly the customers that are interested in low $CO_2$ emissions are also the ones that plan to buy a green car. But it is also shown that not all of them have this plan.

In the following the question was tried to be answered how the development of the customers' attitude towards buying vehicles with a green supply chain is. Fig. A.8 shows that for the majority of the customers it is important that a manufacturer pursues sustainable development strategies.

The customers were also asked how important it is for them that the entire lifecycle of their vehicle is sustainable and negative effects on the environment are reduced. The majority states that it is important or very important to them. This result is a valuable information to the automotive companies. It shows them that their investment in sustainable development of the supply chain is wanted by the customers (Fig. A.9).

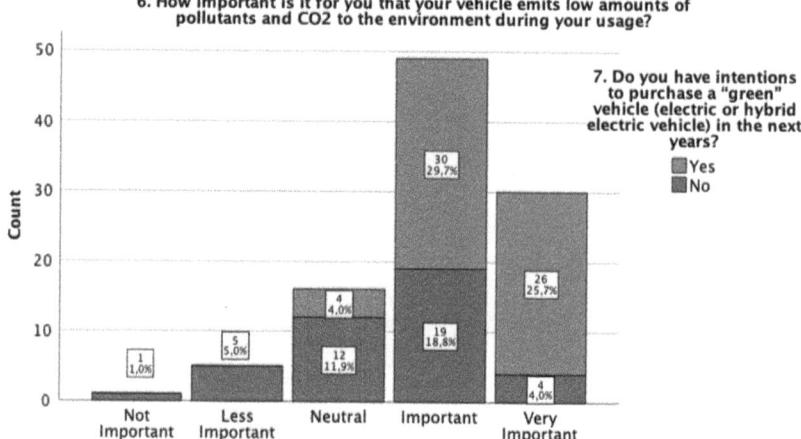

**Fig. A.7** Results of question six stacked by results of question seven

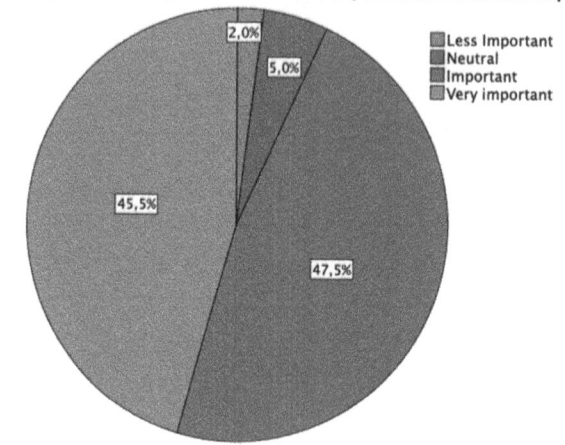

**Fig. A.8** Results of question nine

Subsequentially, it was researched whether that interest in a sustainable or green supply chain is growing compared to the past. In Fig. A.10, it is shown that customers' interest in the green supply chain is indeed growing. Assuming that this trend is going on in the future, OEMs need to expect that the interest of consumers in SSCM will become more and more important to consider in their future development strategies.

10. How important is it for you that the entire lifecycle of your vehicle is sustainable and negative effects on the environment are reduced?

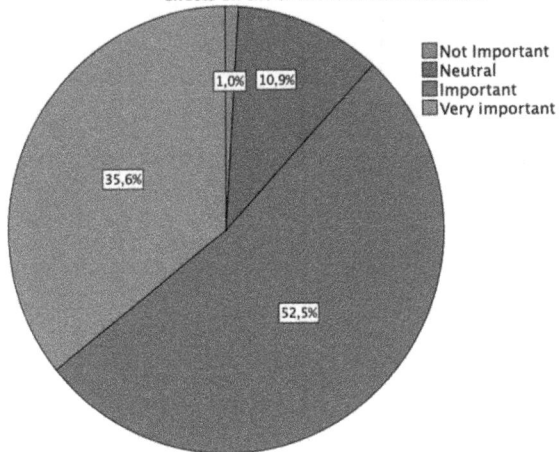

**Fig. A.9** Results of question 10

11. Did your interest in the sustainable supply chain management increase within the last years?

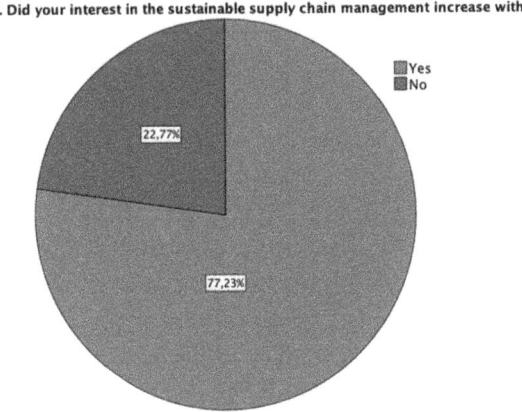

**Fig. A.10** Results of question 11

Figure A.11 focuses especially on the group of customers which are interested in buying a green car in the future. It supports the assumption that this group's concern in the sustainability is higher compared to the other groups. The amount of green vehicle buyers increases towards the

# Appendix

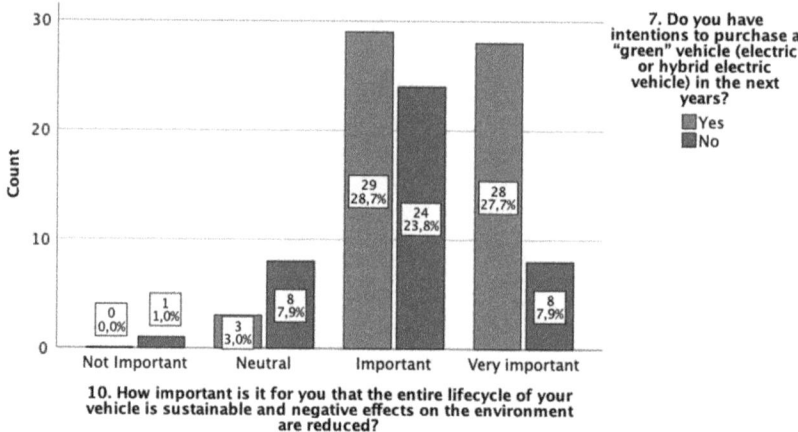

**Fig. A.11** Results of question 10 clustered by results of question seven

right side of the diagram, which represents the customers who answered that SSCM of their vehicle is "important" or even "very important" to them.

Especially in front of the background of the many electric vehicle campaigns with many upcoming electric models announced by the OEMs, it is a valuable information for them, that possible buyers of these cars have an enhanced interest in the sustainability of the supply chain.

The following diagrams deal with the question whether the implementation of SSCM creates value for OEMs. Therefore, the participants were asked if their appreciation and trust in a company would rise if they learned that they care about environmental protection in the entire supply chain. Fig. A.12 shows the very clear trend that the customers agree with this question. Trust of the customer could lead to a higher loyalty to their brands which would result in higher sales and therefore could create value for the company.

Another interesting question is how likely they will accept to pay a higher price for a vehicle that was developed using sustainable supply chain management than for a vehicle developed without it. The majority of the answers state that customers are ready to pay a higher price for vehicles that were developed using SSCM (see Fig. A.13). This information can be used by OEMs for their financial planning. Investments in SSCM could be paid back by higher profits from vehicle sales.

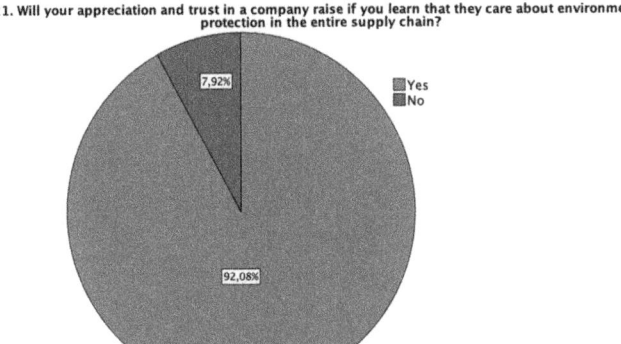

**Fig. A.12** Results of question 21

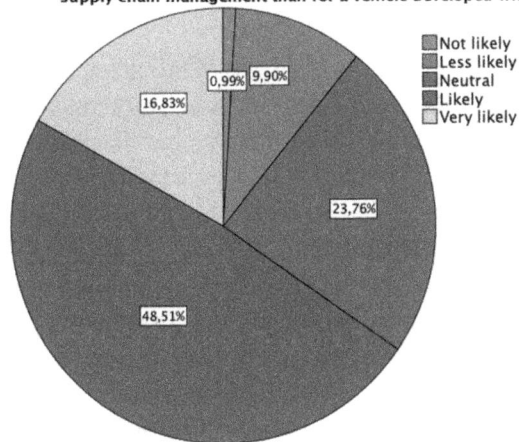

**Fig. A.13** Results of question 17

Lastly, it was examined if customers who research the sustainability of a potential vehicle are also the customers who would accept a higher price for it. Therefore, the participants are differentiated into two groups. They were asked whether they research about the sustainability of the supply chain of a vehicle before they buy it (e.g. green materials, green suppliers,

# Appendix

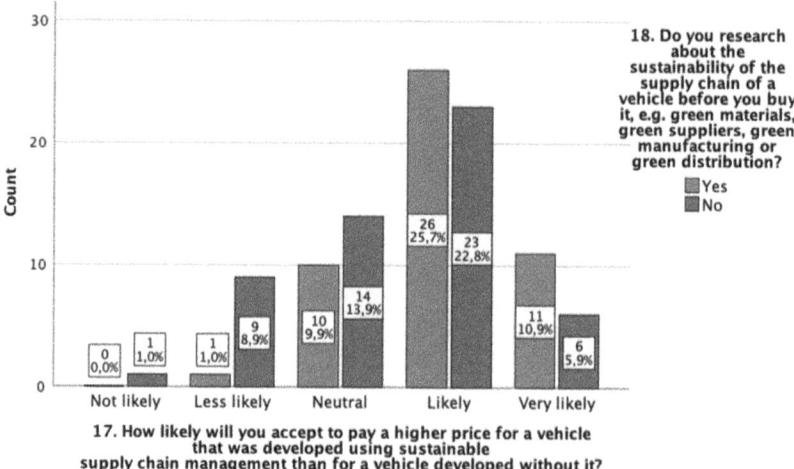

**Fig. A.14** Results of question 17 clustered by results of question 18

green manufacturing or green distribution). One group are customers that answered this question with "Yes" and the other are the ones that answered "No". Fig. A.14 shows the results in a bar diagram that is clustered into the two groups. It shows that customers who would "likely" or "very likely" accept a higher price also show a higher percentage of people that research about the sustainability.

This shows that customers that research the sustainability of the supply chain management of their possible vehicle also accept a higher price for it. OEMs could use this information for future improvements in their communications. Making it easier for customers to research supply chain related sustainability topics could lead to more customers that actually do it and then even start accepting higher vehicle prices after finding out about a sustainable supply chain.

GPSR Compliance

The European Union's (EU) General Product Safety Regulation (GPSR) is a set of rules that requires consumer products to be safe and our obligations to ensure this.

If you have any concerns about our products, you can contact us on

ProductSafety@springernature.com

In case Publisher is established outside the EU, the EU authorized representative is:

Springer Nature Customer Service Center GmbH
Europaplatz 3
69115 Heidelberg, Germany

www.ingramcontent.com/pod-product-compliance
Lightning Source LLC
LaVergne TN
LVHW041204250326
**834689LV00001BA/4**